BROADVIEW LIBRARY

NO LONGER PROPERTY OF
SEATTLE PUBLIC LIBRARY

THE
GREATEST
POLAR
EXPEDITION
OF ALL TIME

the ARCTIC MISSION to the
EPICENTER of CLIMATE CHANGE

MARKUS REX

In collaboration with **MARLENE GÖRING**
Translated by **SARAH PYBUS**

DAVID SUZUKI INSTITUTE

GREYSTONE BOOKS
Vancouver/Berkeley/London

First published in English by Greystone Books in 2022
Originally published in German as *Eingefroren am Nordpol*,
Markus Rex in collaboration with Marlene Göring, © 2020 by
C. Bertelsmann Verlag, a division of Penguin Random House
Verlagsgruppe GmbH, München, Germany
English translation copyright © 2022 by Sarah Pybus

22 23 24 25 26   5 4 3 2 1

All rights reserved. No part of this book may be reproduced, stored in a
retrieval system or transmitted, in any form or by any means, without the
prior written consent of the publisher or a license from The Canadian
Copyright Licensing Agency (Access Copyright). For a copyright license,
visit accesscopyright.ca or call toll free to 1-800-893-5777.

Greystone Books Ltd.
greystonebooks.com
David Suzuki Institute
davidsuzukiinstitute.org

Cataloguing data available from Library and Archives Canada
ISBN 978-1-77164-948-3 (cloth)
ISBN 978-1-77164-949-0 (epub)

Copy editing by Paula Ayer
Proofreading by Dawn Loewen
Indexing by Stephen Ullstrom
Jacket and interior design by Fiona Siu
Jacket photograph by Markus Rex

Printed and bound in Canada on FSC® certified paper at Friesens.
The FSC® label means that materials used for the product have been
responsibly sourced.

Greystone Books gratefully acknowledges the Musqueam, Squamish, and
Tsleil-Waututh peoples on whose land our Vancouver head office is located.

Greystone Books thanks the Canada Council for the Arts, the British
Columbia Arts Council, the Province of British Columbia through the
Book Publishing Tax Credit, and the Government of Canada for
supporting our publishing activities.

The translation of this work was supported by a grant from the
Goethe-Institut.

For Friederike, Tim, and Philipp

CONTENTS

▶ **Translator's Note**

The measurements in this book have been converted to customary units from the original metric and rounded where applicable. For the reader's convenience, temperatures are given in both Celsius (°C) and Fahrenheit (°F).

Prudhoe Bay

150°

180°

*East Siberian
Sea*

Mackenzie   Inuvik

Beaufort Sea

CANADA

120°

*Banks
Island*

*Victoria
Island*

Cambridge Bay

Start of drift

Queen Elizabeth Islands

Resolute

90°W

NORTH POLE

★ *Dranitsyn*
12/13/201

*Ellesmere
Island*

Alert

Pond Inlet

Qaanaaq
(Thule)

Baffin Island

Baffin Bay

60°

★ *Dranitsyn*
2/28/2020

Station Nord

8/10/2020
★ *Tryoshnikov*

*Svalbard
(Spitsbergen)*

Longyearbye

Sonne & Merian
6/4/2020

Fram Strait

GREENLAND

Bjørnøya

30°

Greenland Sea

0°

Ittoqqortoormiit

Qaqortoq
(Julianehåb)

to Bremerhaven
(arrival 10/12/2020)

150°

●Verkhoyansk

Lena

●Tiksi

New Siberian Islands

120°

Laptev Sea

········· *Polarstern active journey*

─── *Polarstern drift*

Pack ice situation on 9/20/2019

Pack ice situation on 3/4/2020

○ Supplied by icebreakers
   or partner ships

Vilkitsky Strait

Cape
Chelyuskin

S I B E R I A

Arctic
Cape

Severnaya Zemlya

RUSSIA

Ushakov Island

Kara Sea

90°E

●Norilsk

Yenisey

Wiese Island

Dikson●

80

*Shokalsky
Island*

*Franz Josef
Land*

Cape
Zhelaniya

Novaya Zemlya   60°   70

Arctic Circle

Polarstern route

Vaygach
Island

Ob

Kara Strait

Barents Sea

60

Pechora

N O R W A Y 30°

North Cape

9/20/2019
● **Polarstern launches**
Tromsø

●Murmansk

●Archangelsk

Dvina

500 km
(310 mi.)

0

PROLOGUE

Unseen and untrodden under their spotless mantle of ice the rigid polar regions slept the profound sleep of death from the earliest dawn of time. Wrapped in his white shroud, the mighty giant stretched his clammy ice-limbs abroad, and dreamed his age-long dreams.

Ages passed—deep was the silence.

Then, in the dawn of history, far away in the south, the awakening spirit of man reared its head on high and gazed over the earth. To the south it encountered warmth, to the north, cold; and behind the boundaries of the unknown it placed in imagination the twin kingdoms of consuming heat and of deadly cold.

But the limits of the unknown had to recede step by step before the ever-increasing yearning after light and knowledge of the human mind, till they made a stand in the north at the threshold of Nature's great Ice Temple of the polar regions with their endless silence.

Up to this point no insuperable obstacles had opposed the progress of the advancing hosts, which confidently proceeded on their way. But here the ramparts of ice and the long darkness of winter brought them to bay.[1]

THUS BEGINS FRIDTJOF NANSEN'S 1897 account of his pioneering polar expedition. Since then, we have unraveled most of our home planet's secrets, studying and measuring even the most remote areas with the latest scientific instruments. But the polar regions continue to thwart our urge to explore. Even today, the light of our knowledge dims as the central Arctic, the region north of the eighty-fifth parallel, enters its winter months. The ice on the Arctic Ocean is too impenetrable, the conditions too adverse. No research icebreaker has managed to find a way through. No one has studied the central Arctic's complex climate processes all year round.

Until now. The brainchild of twenty nations, the MOSAiC expedition (Multidisciplinary drifting Observatory for the Study of Arctic Climate) is a major feat of international collaboration. Aboard the *Polarstern*, the first modern research icebreaker to spend an entire year in the central Arctic, a team of researchers reveals the secrets of the Arctic and studies the North Pole's immediate surroundings, even in winter. In an expedition that pushes the boundaries of possibility, the *Polarstern* spends the winter firmly frozen into the central Arctic ice, supported by a fleet of six further icebreakers and research vessels, devoting an entire year to collecting the data we so desperately need.

The Arctic is the epicenter of climate change. It is warming more rapidly than anywhere else on Earth, at least twice as fast as the rest of the planet and even faster in winter. There is still much we do not understand. The Arctic is where our climate models operate with least certainty; warming predictions to the end of the century can differ by up to a factor of three—in the worst-case scenario for future greenhouse gas emissions, temperatures are forecast to rise by between 5°C (9°F) and a whopping 15°C (27°F). Many models predict that in just a few decades, the Arctic will have no ice in summer. Other models do not. Nobody knows whether or when this will happen. But society needs robust and reliable scientific foundations on which to base urgent, far-reaching climate-protection measures.

Climate models are based on data and a precise understanding of climate system processes, which we must re-create as realistically as possible with our computers. This is the only way for models to return reliable results. But how can we do this for a region in which we have never managed to observe the climate system with our state-of-the-art equipment? Without these observations, models have to make ad hoc assumptions about how these processes work. You might say they have to guess. This leads to huge uncertainties in their projections of climate change.

And yet the Arctic is the crucible of weather and climate in Europe, North America, and Asia, which together house a significant proportion of the world's population. The contrast in temperatures between the cold Arctic and the warmer mid-latitudes drives the main wind system in the northern hemisphere and plays a considerable role in our weather. The rapid warming of the Arctic alters this contrast, increasing and intensifying extreme weather in our latitudes. With so little information about Arctic processes, it is currently difficult to say with any certainty what an ice-free Arctic summer would mean for our climate.

So how do you reach the central Arctic in winter, when the ice is so thick that even our best research icebreakers can't get through? Our expedition follows in the footsteps of Fridtjof Nansen, the polar pioneer who discovered the Arctic ice drift. Inspired by the discovery, north of Greenland, of the remains of the 1879 *Jeannette* expedition, which foundered in the ice off Siberia, Nansen used the Transpolar Drift—a "conveyor belt" of ice that moves across the polar cap—to travel deeper into the central Arctic than ever before. He allowed the *Fram*, a specially built wooden sailing ship, to be trapped in the ice off the Siberian coast in the area where the drift begins; within three years, the ice flow had carried the ship across the polar cap and into the North Atlantic.

The MOSAiC expedition adopts the same approach. We work with the ice, not against it. The idea is that, if we allow ourselves to be trapped in the right part of the ice, the Transpolar Drift will carry us straight through the central Arctic with no further action on our part, granting us access to this normally isolated region—even in winter. We will spend the long winter and spring period firmly and inescapably embedded in the drifting ice.

We are entirely at the mercy of natural forces; nobody can say or influence where the ice drift will take us or how the expedition will go.

We are relying on nature. We do not determine our course. On a venture such as this, we cannot make plans; we must react to events as they unfold. Great challenges await us—fissures in the ice, mighty peaks forming as ice sheets dramatically collide, violent storms, extreme cold, the impenetrable blackness of the months-long polar night, dangerous polar bears, and, of course, the coronavirus pandemic. We are ready.

Part I

FALL

▶ The *Polarstern* heads into the ice.

1

IT BEGINS

September 20, 2019: Day 1

THE *POLARSTERN* SITS at Tromsø pier, resplendent, her hull illuminated in the encroaching darkness. It's time! I stand on board and look at the crowd celebrating on the shore. Our starboard decks—our landward side—are full of people. Our team of around one hundred scientists, technicians, and crew members is embarking on an adventure that will see us frozen into the ice for months on end, alone at the end of the world, on the biggest Arctic expedition of all time.

I look down. To commemorate our departure, artists have projected a light installation—a moving ice floe   onto the concrete of the pier; in the shipyard, the party tent glows in the dusk. Anja Karliczek (German minister of education and research), Otmar Wiestler (president of the Helmholtz Association), and Antje Boetius (director of the Alfred Wegener Institute, or AWI) have given speeches to send us on our way. This is an honor for us and our project—and a sign of just how important the Arctic and global warming are now considered by politicians and society. The press has turned out in force. Before boarding, we drank a toast and said a special goodbye to Antje, who has advocated for our expedition so fiercely for so many years, helped with the planning, and made so many things happen; we know she would love to be traveling with us.

Is that a teardrop I can see? It must be the brisk Tromsø wind! Uwe Nixdorf, head of our logistics department, is there along with Klaus Dethloff, who birthed the idea for MOSAiC years ago. They watch with pride as our plans become reality. An expedition like ours cannot be achieved by one country or one institution alone; many people have been working and fighting for this for a long time—including Matthew Shupe, standing next to me on board. We are all thrilled to see our efforts pay off. The guests raise their glasses; the *Polarstern*'s hull puts several arm lengths between us. Friends and family wave from below, my wife and two sons among them. We wave back. Many pairs of eyes seek out their loved ones for one last glance. But the atmosphere is too exuberant for tears; this is no place for melancholy thoughts.

And now it's time to go! The band plays, the gangway lifts, we cast off, and with a long toot of her horn, the *Polarstern* gently starts to move. Soon we can no longer make out the people in the harbor. Our friends disappear into the darkness and the music is swallowed by the wind.

I stay on deck for quite a while, looking out over the fjord. The lights on the coast, and later the Norwegian islands, pass by in the darkness. Cozy lights in snug Norwegian houses. Just like the little house I share with my family in Germany, in the heart of Potsdam-Babelsberg, a district not far from Berlin that feels almost like a village. It will be a long time before I see it again. While the people on the shore—and their families—are coming to the end of a normal day, our team will not have a normal day or see our families for quite some time. What will the coming months bring?

As I think back, the months leading up to the expedition seem almost surreal, a whirlwind of final preparations. My house looked like an expedition camp, mountains of things waiting to be packed. But more than anything, the time I had left with my family became more precious as each day passed. Slowly it sank in that in the coming year, we would spend nine whole months apart, including Christmas, New Year's, and all our birthdays. Despite this, my two sons—aged nine and eleven—are thrilled by the expedition. They know all about it and share our excitement. This comforts me somewhat in the face of our long separation. My wife has never known me otherwise—I have always embarked on long expeditions—and is equally enthusiastic. At least we can send each other messages; previous generations of polar explorers couldn't even do that.

During this time, I often think about Fridtjof Nansen and his team, who embarked on a very similar voyage 126 years ago and demonstrated that such an expedition was indeed possible—an enormous achievement given their ship, the *Fram*, was made of wood. They set off into the complete unknown, with no way to communicate with the outside world and no idea of whether they would ever come back alive. How did they feel in the final days before departure? What anxieties must have plagued those men (yes, back then they were all men; things are different now) and their families? And how much better do we have things today?

▶ **THE *POLARSTERN***

The *Polarstern* has been sailing to the most remote corners of the planet since 1982. She has many roles: she supplies the German research station in the Antarctic (Neumayer Station III on the Ekström Ice Shelf near Atka Bay) and is used almost constantly in the polar regions to research ice, the ocean and the life within it, biogeochemical processes, the atmosphere, and the climate. On average, the *Polarstern* spends 310 days of the year away from her home in Bremerhaven. She is also one of the world's most capable research icebreakers; she has a strong double hull with a typically rounded shape that can easily break through ice five feet thick. Her 20,000 horsepower provides enough force to ram through even thicker ice. She is a floating research institute, housing nine laboratories with highly specialized instruments. New features added for MOSAiC include a scientific container deck in the bow.

Now we are actually on our way! The *Polarstern* creates foamy waves at first, then her wake grows smaller, merges with the ocean, and disappears entirely. It feels good to watch, a sign that we are detaching ourselves from all those years of planning and the final, stressful days in Tromsø harbor.

Now, the overwhelming pressures of the last few days ashore give way to the calm that always descends as a ship glides slowly and steadily through the ocean—particularly as the wake disappears into the dark nothingness of the nighttime sea.

Slowly, it dawns on me that we are on our way, that we have nobody to rely on but ourselves. For better or worse, we have only what we carry with us; this goes for our socks, headlamps, and woolly hats as well as our equipment and provisions. We can't stop off to do some shopping. There won't be any deliveries. We can't expect any help from the outside world.

Paradoxically, this is reassuring. The world has suddenly become very small. We have few options or courses of action, but even that is strangely relaxing. There's no point in frantically listing last-minute tasks or purchases. Before sailing, we began to think in ever-smaller units of time, ultimately in hours and minutes, but now we have all the time in the world. The expedition will last a year. It's a marathon, not a sprint. And that is something to be approached with peace and serenity. I unpack a few of my many boxes, then go to bed. Within a minute I'm sleeping like a log. You can always rely on the *Polarstern* to rock you to sleep.

September 21, 2019: Day 2

OUR FIRST MORNING at sea. We can still see a few Norwegian islands on the horizon, gleaming through the overcast sky, but by midday they have disappeared completely. We also lose our cell signal. Civilization's radio waves can't reach us now. The *Polarstern* fearlessly pitches through the churning water. We round Scandinavia's North Cape at around 11:00 AM and follow a northeast course, into the Barents Sea, which is open and ice-free in the late summer.

It's good to feel the movement, the familiar rolling and pitching, as the *Polarstern* forges ahead. I'm drawn to the P-deck, the highest deck above the bridge, the brisk wind blowing around me, the pitching ship below, the unencumbered view all the way to the horizon. This is one of my favorite spots on the vessel.

September 22, 2019: Day 3

WE ARE MAKING excellent progress through the open water of the Northeast Passage, traveling at a good thirteen knots (15 mph) against the wind. The wind grows stronger as the day goes on, and the *Polarstern* continues blithely through waves that reach an average height of thirteen feet. Occasionally the water washes over our working deck, and some people start to feel seasick. But the mood on board remains excellent. After all those years of planning, we are enthusiastic and can't wait to reach the ice.

We have made ourselves at home and stowed our luggage in our cabin lockers; along the corridors, each door has a row of bulky work boots and padded polar boots to be worn on the ice. There are two people per cabin, each of which features a bunk bed, a little seating area with a table, a small separate bathroom, and not much else. My single cabin consists of a bedroom and an office with a cozy seating area.

But we don't spend much time in our cabins. We're already working all day. The laboratories have to be set up, the boxes unpacked, and the instruments calibrated. Despite having a lot less space than on land, we rack up huge step counts between the decks, labs, and containers.

The *Akademik Fedorov*, our escort ship for the first leg of the journey, left Tromsø yesterday afternoon. She was supposed to cast off with us but had to wait for equipment that arrived at the harbor too late. Now the *Fedorov*, the flagship of the Russian polar research fleet, is following our route into the ice. She's carrying additional equipment and people

▶ WORKING DECK

The main point of access to the ice via a gangway. In a heated cabin, the "gangway watch" record who is on the ice and who has returned. The back of the deck has another heated cabin with a semicircular panoramic window, where the "stern watch" look out for polar bears and secure the area behind the ship that can't be seen from the bridge.

▶ HELIDECK

This is where the *Polarstern*'s two BK117 helicopters take off and land when running scientific missions or providing an overview of the ice conditions. This is also where we release research balloons into the stratosphere.

▶ LABORATORY CONTAINER

The ship's hold houses several laboratory containers, both here and at the bow. Some can be cooled to various low temperatures, while some provide specific light conditions for biological work. Other containers are filled with devices that take atmospheric measurements, drawing in air from outside the ship.

▶ SLIDE BEAM

In the fourth and fifth expedition phases, this is used to lower the large, heavy ocean instruments deep into the water column through a hole in the ice. In the first and second phases, a new procedure is used involving the ship's crane and a hole in the ice farther from the ship.

▶ LARGE WET LAB

The largest workroom on the expedition. This is where the large remote sensing instruments are set up. It is also the home of the drones and the HELIPOD, a towed system for atmospheric measurements that is flown with a helicopter.

▶ CROW'S NEST

This has a rotating infrared camera that keeps a constant 360-degree watch for polar bears—even in the total darkness of the polar night. However, it breaks during the first expedition phase, as does the identical replacement camera. We are then left with two reliable infrared cameras: one that can swivel and zoom, and one that continuously monitors the area behind the ship.

▶ P-DECK/MONKEY DECK

This is the location of the satellite antennae for data transmission. Atmospheric scientists use the unobstructed view of the sky above and have installed (among other things) a large, swiveling cloud radar; this is so heavy that the P-deck had to be specially reinforced.

▶ BRIDGE

The command center for all activities. Work in the ice camp is constantly monitored from the bridge. It is the central unit for radio communications with the teams on the ice and serves as a permanent lookout point for polar bears around the camp. During winter's polar night, we use the infrared cameras (which are controlled from here) along with our three searchlights; in summer's polar day, we use binoculars.

▶ MEETING ROOM

The expedition team meets here at least once a day. Polar bear shifts are arranged for the following day and volunteers are often sought—and always quickly found—for a range of tasks in the research camp.

▶ LAB ROOMS

Many of the water samples obtained from the depths of the ocean using the CTD rosette (page 49) are analyzed here or secured for analysis in laboratories back home.

▶ FORESHIP

A whole new deck has been constructed on the bow of the *Polarstern* for MOSAiC with measurement and laboratory containers. Countless atmospheric measurements are taken here throughout the expedition.

who will help us set up our research camp, as well as the network of observing stations distributed on smaller ice floes around our base, up to thirty miles away. Before docking at our ice floe, she will bunker the *Polarstern* to replace the fuel we have used so far, allowing us to enter the long winter with full tanks.

Our next goal is to circumnavigate Cape Chelyuskin, the northern-most point on the Afro-Eurasian landmass and the crucial section of the Northeast Passage. When we pass the Cape, we will have the Laptev Sea before us; somewhere north of this, we aim to be trapped in the ice. But first, we need to cross the rest of the Barents Sea and the Kara Sea.

There are two ways to do this. The ice might force us to stay near the coast and hope for open passages. This route would take us through the Kara Strait—the narrow gateway to the Kara Sea—between Novaya Zemlya and Vaygach Island, near the mainland. Or we could travel north around Novaya Zemlya and fight our way eastward through the northern Kara Sea. The ice will make the decision for us.

The Kara Sea's tricky ice conditions have earned it the German nickname *Eiskeller* (ice cellar), coined in the mid-nineteenth century by Karl Ernst von Baer, a Baltic-German scientist and explorer. But there's no sign of an ice cellar here; the Kara Sea is almost completely free of ice and stands open before us! So we take the easier and quicker route and head for Cape Zhelaniya, the northern tip of Novaya Zemlya. What a difference from the era of Fridtjof Nansen, the inspiration for our expedition!

Fridtjof Nansen: Discovering the Drift

OUR EXPEDITION IS modeled on Fridtjof Nansen's endeavors between 1893 and 1896. Nansen discovered the natural drift of the ice and was the first to use it to carry his ship, which is precisely what we intend to do. He traveled deeper into the Arctic than ever before, at a time when some thought the North Pole might lie in an ice-free ocean or even on an undiscovered continent.

▶ Fridtjof Nansen discovered the Transpolar Drift (thick arrow), which is part of the Arctic's natural ice drift. The top-left arrows represent the Beaufort Gyre, while the shading indicates the typical ice spread in the summer.

For centuries, people have bravely tried to forge a path through the Arctic Ocean. The great unknown beyond the ice edge fires the imagination. The urge to discover what lies in this vast region has claimed the lives of many explorers. But not Nansen of Norway. He set sail on the *Fram*, his three-masted ship, with a team of just thirteen. With a strong outer layer, a rounded hull, and an enormously stable internal design never seen before, the ship's structure ensured that the pack ice would not crush it but simply lift it; even the rudder was retractable.

Five years previously, aged twenty-seven, Nansen had crossed the Greenland ice sheet on skis with just four companions, learning from the Inuit how best to survive in the Arctic. The sleds we are using on our expedition look essentially the same as the ones Nansen had built for the *Fram* expedition. Based on what he saw the Inuit use, they are flat to distribute their load, with movable struts that prevent the sled from breaking when pulled over the rough ice. Nansen's provisions included a large quantity of dried fruit, which protected his crew from the dreaded scurvy, even though the link between the disease and lack of vitamins had not yet been identified precisely.

▶ Fridtjof Nansen (middle) measures the solar eclipse with Hjalmar Johansen and Sigurd Scott Hansen. Polar region, April 1894.

▶ This famous portrait of Nansen was taken by Henry Van der Weyde in the 1890s.

All of these measures were designed to aid his plans. Nansen wanted to be the first to travel through the Arctic Ocean with the ice, not against it. He deliberately sailed into the pack ice off Siberia in order to be carried across the North Pole and back to Greenland. Nansen got the idea from a pair of oilskin trousers found in 1884 on an ice floe off Qaqortoq (formerly Julianehåb) on Greenland's coast, along with further remains of the *Jeannette* expedition. But the *Jeannette* failed in her mission to travel deep into the Arctic from California via the Bering Strait; she was trapped and crushed by the ice in the East Siberian Sea. So how did her wreckage get from Siberia to Greenland? Nansen deduced that there must be a natural ice drift across the Arctic that had carried the *Jeannette*'s remains from Siberia to Greenland.

Nansen was right. Sea ice isn't static; it moves through the Arctic Ocean. We know all about this today. The Transpolar Drift travels from the areas north of Siberia, through the central Arctic and into the Arctic's Atlantic sector. The drift forks north of Greenland; one part turns off into the Fram Strait (where Nansen's *Fram* was spat out by the ice, and which therefore bears its name) while the other continues into the Beaufort Gyre, where sea ice turns clockwise off the coasts of Greenland, Canada, and Alaska.

Many of Nansen's contemporaries thought he was crazy—and accused him of irresponsibility—for voluntarily sailing a ship into the pack ice. But Nansen wouldn't be deterred. On June 24, 1893, he set off from Christiania (now Oslo); three years later, the team ended their expedition in Tromsø, where our voyage began. Everyone on board was safe and sound, but the *Fram* arrived without Nansen or his companion, Hjalmar Johansen. While the ship continued to drift in the Arctic Ocean from the northern Laptev Sea to Spitsbergen, Nansen and Johansen spent their second spring making for the North Pole on skis and sleds. They may not have reached it, but they did set a record for attaining the highest northern latitude at that time. Nansen and Johansen also returned safely from the ice at almost the same time as the *Fram*. This was another milestone—it would remain the only wooden ship to travel so far north.

Most notably, Nansen provided countless Arctic insights without which our work today would be impossible, and he inspired our own expedition on the *Polarstern*. What makes him so fascinating is that he wasn't just a man of action; he was thoughtful, almost melancholy. He didn't possess the hubris of other explorers, some of whom believed that the Arctic's Indigenous inhabitants had nothing to teach them—which is why they used horse-drawn sleighs rather than teams of dogs and preferred to load their ships with silverware rather than kayaks, the usual method of rescue from the Arctic ice. This attitude has led many explorers to their death, both before and after Nansen. Nansen was humbled by the unrelenting force of the polar regions. His was not a heroic attempt to conquer nature. He was successful precisely because he patiently adapted to its conditions and learned from Inuit how to survive in this forbidding environment. His expedition report details his routes and adventures, but also contains reflections on nature and our role within it. After his time as a polar explorer, he pursued a second career as a diplomat and was awarded the Nobel Peace Prize for his work on behalf of refugees in World War I. A truly formidable character.

I often read Nansen's records from the *Fram* and have brought the two huge volumes along with me. As we follow almost the same route that he took 126 years ago, I can see how much the world has changed. In the dreaded Kara Sea, Nansen struggled along the Siberian coast, obstructed again and again by ice fields that stretched from the central Arctic to the coastline. But our path is clear; in late summer 2019, the Kara Sea doesn't have enough ice for so much as a glass of whiskey.

Things are a little different when I look ahead on the satellite images, through the northern Kara Sea and farther along the traditional Northeast Passage; the direct point of access to the Laptev Sea north of Severnaya Zemlya is blocked by a tongue of ice located east of the islands and continuing south. The island group includes the last large islands to be discovered on our planet. Located off Cape Chelyuskin, they are not shown on Nansen's maps. At that time they were so ensconced in ice that Nansen must have sailed right past the cape without even noticing the large and mostly glaciated islands to the north.

Now we have to figure out the best way to our destination in the current ice conditions. Should we take the ice-free but longer route through the Vilkitsky Strait, which would lead us close by Cape

Chelyuskin between islands and mainland—or the direct route past the northernmost point of Severnaya Zemlya, the Arctic Cape, and then try to forge a path through the ice? Or should we maneuver between the islands and across the Shokalsky Strait, a narrow, rocky passage, to avoid at least most of the ice tongue?

To assess the situation, you need to know the thickness and stability of the ice. The satellite images tell us nothing about that, but, as luck would have it, two ships visited the area late this summer: The *Akademik Tryoshnikov*, an icebreaker belonging to the Russian colleagues who brought our fuel depot out to Severnaya Zemlya (more on that later) and the *Bremen* from Germany. We gather information about the regional conditions from the two ships. Both report that the Shokalsky Strait is currently navigable but blocked on its eastern side by several stranded icebergs that must be negotiated carefully—no easy undertaking in the strait's frequent and dense fog.

The *Tryoshnikov* tells us that the ice tongues are made of solid ice up to five feet thick in places, and that it would be better to go around them. The captain and I agree to keep all three options open. For the time being, we remain on course for Cape Chelyuskin.

September 23, 2019: Day 4

WE ARE SLOWLY settling into a routine. We spend as much time as possible out on deck, watching as we glide through the waves, enjoying the movement and the visible progress now that we are setting our own course. Things will change soon. Once we allow ourselves to be trapped in the ice, nothing will move; everything will become static and our route will be determined by the ice.

And yet everyone is clearly excited to reach the ice. This will be the start of the main expedition phase, the one we all came for: drifting through the Arctic with the ice. We can't wait.

In the evening we open the bar for the first time; it's been lovingly dubbed the Zillertal thanks to its acquired-taste décor reminiscent of the mountain huts of Austria's Ziller Valley: tablecloths, red lights, and a counter adorned with logo stickers from previous voyages. The atmosphere is wonderful, and the expedition members are really getting to know one another. We are gradually becoming a tight-knit group, a little bunch of people who will spend the next few months together in the central Arctic ice, hundreds of miles away from the next human soul. We know that magnificent experiences await us, and that forges close bonds very quickly.

▶ The team for the expedition's first phase in the *Polarstern* meeting room.

Later that evening, the Arctic greets us with one of its most impressive spectacles—the aurora borealis. A broad arc of green light paints the sky. It moves languidly against a backdrop of twinkling stars, like a curtain rippling gently in a breeze. One section of sky suddenly erupts, unraveling in a spiral, growing brighter and trickling down, sending fingers of light into the firmament, then slowly growing calmer, fading out. A few minutes later, the next flash of light appears elsewhere, a new wave crashing across the sky, twisting and waning once again. While the ribbon of light ripples across the zenith, star-shaped beams descend, green, shimmering slightly in shades of red to violet at their highest point. I lie on the P-deck for hours, watching the display as the ship plows on.

The aurora never fails to fascinate me. I've seen it so many times, but until now I've always associated it with bitterly cold air, snow crunching beneath my feet, and my breath visible in front of me—exactly how the high Arctic feels in the depths of winter. In the summer, the polar day is too bright to see the northern lights.

I can still remember the first time I saw the aurora, in Spitsbergen in January 1992. Polar night. Back then I didn't know the Arctic at all, and with the permanent darkness, I had no idea how the region looked in daylight. Nevertheless, I moved far away from our research station so that I could watch the spectacle with no interference from its lights—armed with a gun, of course, in case of polar bears. Everything was new to me: the icy Arctic air that you can feel on every patch of exposed face, but that is too dry to freeze. And then the sounds that snowy surfaces make below −30°C (−22°F). If the top layer of snow, compacted by the wind, is subjected to load, it can sometimes break several yards away and then emit strange noises. Out there in the darkness, where polar bears might be lurking, it sounds like steps. But I recognized the effect of the cracking snow and reassured myself that it was nothing. Then I heard a creaking and crunching a short ways behind me. Of course it was the fjord the ice makes these sounds as it moves across the water. There were no bears; I could stay and watch the aurora in the singular Arctic air.

I spent hours out there, soaking up the sensations. The aurora isn't even at its most active on Spitsbergen—our research station was a little too far north. In northern Finland, I skied down a frozen river for hours as the lights danced above me, growing more and more intense.

The sky feels incredibly large when the aurora appears. It gains a third dimension and seems more than ever like a huge dome over our planet. The shapes constantly shift—slow and sedate, then fast, but never hectic. The aurora exudes an unbelievable calm.

There are reports, myths really, that the aurora is accompanied by sounds. I've never heard anything and tend to associate the northern lights with the Arctic's absolute tranquility and scentless, icy air.

And now here I lie on the deck of the rolling *Polarstern*, the temperature above freezing as I watch the display unfold. The air is different out here on the ocean, much warmer and full of smells. No two moments are the same; you could spend several lifetimes in the polar regions and experience something new every time—if you're willing to look closely and soak it all in.

Will we now see this phenomenon on every clear night? That was what Nansen's team reported. But the aurora borealis is complex. It occurs when solar wind meets the Earth's atmosphere. Solar wind consists of charged particles—mainly electrons and protons—which are intercepted by the Earth's magnetic field, deflected, and propelled toward the Earth over the polar regions, where they form an oval around the magnetic poles in the upper atmosphere, lighting up. At the same time, the particles alter the magnetic field itself, which is why the aurora moves in so many ways. However, there are two reasons why we are less likely to see many auroras later in our journey—unlike the frequent and spectacular displays on Nansen's expedition.

First, the auroral oval moves around the magnetic pole at a distance of about twenty latitudes—and the magnetic pole migrates over the decades. In Nansen's day it was in northern Canada; in the last few years it has positively sprinted toward (and is now relatively close to) the Geographic North Pole. We will soon be significantly farther north than Nansen and thus much closer to the magnetic pole. And that means we will be *too far north* for the aurora borealis! We will spend most of this expedition on the other side of the aurora, too far away for a good view.

Second, solar wind follows an eleven-year cycle. It was extremely active when Nansen traveled but will be at its lowest during our expedition.

We probably won't get to see many aurora displays later on, so we spend much of the night staring into the sky. Soon we will reach the ice.

2

ON THIN ICE

September 24, 2019: Day 5

WE LEFT NOVAYA ZEMLYA behind in the night and are now traveling northeast through the Kara Sea. We head for the Severnaya Zemlya island group. Beyond it lies the ice edge. We continue to make good progress on the open water, moving at around twelve knots against strong wind. The sea is lively and rocks the *Polarstern* jauntily. The sun barely shows its face, just like every other day.

According to the latest satellite maps, the ice tongue we previously saw on the eastern side of Severnaya Zemlya has retreated somewhat, so we decide to sail north around the islands. Then we'll try to break through the ice on the east to access the open water of the Laptev Sea. We set a course for the Arctic Cape, the northernmost point of Severnaya Zemlya.

The Barents Sea, Kara Sea, and Laptev Sea are shallow marginal seas on the Siberian continental shelf. The water is rarely more than 650 feet deep, and the seabed is often only a couple of hundred feet away on the sonar. But the *Polarstern* has a draft of thirty-six feet, and the charts for this region are deceptive and incomplete.

On our way through the Kara Sea, we pass two little islands, Ushakov and Wiese. Our navigation officer has used the charts to set a course that should lead us safely between them. The water is supposed to be over 500 feet deep here, but suddenly the number on the echo sounder drops rapidly: 300 feet, 250 feet, 200 feet ... just 115 feet! Up on the bridge, the watch officer springs into action. Hard to port, away from Wiese Island on the starboard side. The depth increases rapidly. That was a close call. We're forced to swerve a second time; even today, the shallow Kara Sea is poorly mapped.

September 25, 2019: Day 6

IT'S EARLY MORNING when we round the Arctic Cape at a distance of around twelve miles. The sky is hazy and overcast, so we can't see it. We miss our last glimpse of land. We are now in the Laptev Sea, and it can't be long until we see the ice.

First, we head southeast—south to bypass the ice tongue—then resolutely east, directly toward the Arctic ice. We all wait expectantly for the moment when we reach the ice edge. In the afternoon we crowd onto the deck and bridge, gazing into the distance, mesmerized.

The Arctic has its own, very subtle beauty. It's perhaps not as ostentatious and instantly overwhelming as the Antarctic, with its breathtaking icebergs, superlative ice cover, and the world's largest colonies of bustling penguins and other creatures. You have to surrender yourself to the Arctic. Its beauty lies in its endless icy expanse. In its absolute silence, interrupted only by muted crunching and creaking as the ice pushes and scrapes. The bitterly cold air, the snow crystals floating gently across the vast ice, the extraordinary light that changes, little by little, as the year goes by.

Of course, it can make a much more obvious and dramatic impact: for example, when pressure ridges—massive ridges of compressed ice—emerge with great force, huge polar bears lumber by, or the transcendental northern lights flash across the sky. But to me, the Arctic's understated aspects are what make it so special. They demand your undivided attention, to be absorbed slowly. They are what keep me coming back for more.

Gradually we start to see ice floes drifting past in the open water, small and fragile at first, then larger and in greater numbers. One of the first, to the front on our portside, is carrying a polar bear. He sits placidly and watches the ship with curiosity. As our blue, white, and orange steel colossus draws closer, he gets spooked, jumps into the water, and swims away with powerful strokes.

And then, quite suddenly, at about 3:00 PM, all that we see before us is ice.

The ice edge appears in the sky before we reach it. Above us the sky is gray, but in front it is a bright, starkly contrasting white. This effect is known as "iceblink."

Iceblink occurs because ice reflects most light back onto the clouds—unlike the dark, open sea on which we are sailing. The vast surface of the ice creates an almost inverse image in the sky, making it glow brightly. And it's exactly the same the other way around: when you're deep in the ice, open water on the horizon is indicated by an ominous gray in the cloudy sky, even though the sky is actually white. This is known as "water sky." If you're not familiar with this effect, you might think a violent storm is coming—in fact, it's because the undersides of the clouds lack light if there is no ice below them to reflect it. Seafarers like to use water sky to spot channels of open water through the ice that will make it easier to maneuver their ship. They can identify open routes long before they actually see them.

And then, finally, we reach the ice edge. The ship rears up. The hull shudders as we meet the hard ice floes. And the *Polarstern* does what she does best, pushing on valiantly through the ice. It creaks, cracks, and rumbles, pushes the ship back and forth, but nothing can stop the *Polarstern*. When confronted by a huge ice floe, she pauses before laboring through and continuing on her way.

▶ On September 25, the *Polarstern* reaches the ice edge east of Severnaya Zemlya.

When faced with a thicker ice floe, the ship rides up slightly onto the ice and bears down until it breaks beneath her—icebreakers are built for this, with their flat, bulbous bows. Cracks fan out in front of the *Polarstern* and she sails in, pushing the ice apart. The remains of the broken floes end up vertical beside the ship as she pushes through. Even in the ice, we can often reach speeds of seven knots. The thicker sections force us down to two or three knots, but we rarely come to a standstill.

When we do, we employ the second icebreaking method: ramming. When the *Polarstern*'s propellers are turning, we can quickly divert her power in the opposite direction, changing the gradient of the huge blades to create a backward thrust. The sudden reversal sends vibrations through the ship and she starts to move back down the route we have created in the ice. The helmsman then directs her force forward again and she picks up speed as she moves toward the ice, her enormous mass building momentum—and then she slams into the ice. Ideally, the impact will break the ice and clear the way, but sometimes the maneuver has to be repeated several times before the ice yields.

The ship moves steadily and more predictably on the open sea. You can tell more or less which way she's going to tilt next. If the sea is rough and you're trying to climb the stairs, you quickly realize that it might be best to wait for the waves to help you, rather than fighting the up-and-down motion. After a while at sea, your movement patterns are basically programmed to match those of the ship.

But now the *Polarstern* is traveling through ice, and this feels very different.

The ship could tilt at any time and without warning, even after many minutes of smooth sailing. Riding up onto an ice floe can send her to one side; the movement comes out of the blue, and often sends a shock wave through the whole ship. To our surprise, the tongue of ice we are now crossing on our way through the Laptev Sea contains massive, hard chunks of ice over six feet thick. They must have formed during powerful ice compressions off Severnaya Zemlya, where the ice accumulates. When we run into one of these chunks, there's a loud bang at the bow; the *Polarstern* throws herself to the side and crushes the ice with her weight and momentum. The experience is particularly memorable in the sauna, which is located in the bow, next to the hull, close to the waterline. Ice is constantly cracking and sliding along the hull. While you sit there and sweat, thick chunks of ice smash right past you. The sauna's always busy when we travel through ice.

September 26, 2019: Day 7

WE BROKE THROUGH the hardest part of the ice tongue overnight and are now traveling through loose drift ice, alternating between ice fields and stretches of open water as we continue east. With frequent gaps in the sea ice, this is a good place to calibrate our magnetometer. We spend the morning sailing in a clean figure eight: two circles each two miles across. We cross every direction of the Earth's magnetic field and continue our voyage with a freshly calibrated magnetometer. This will allow us to precisely measure the magnetic field throughout our drift.

September 27, 2019: Day 8

WE HAVE LEFT the shallow shelf sea behind and entered the central Arctic Ocean basin. The water here is largely about 10,000 to 13,000 feet deep, but right now we are directly above an impressive valley with a central depth of around 18,000 feet—the Gakkel Deep, one of the deepest points in the Arctic Ocean.

We won't go any farther today. We spend the day recovering four instruments that were installed on the seabed by a previous expedition. It all goes smoothly; by the evening three of the four are on deck waiting to be returned on the *Akademik Fedorov*, our escort ship.

We put the break to good use. A few colleagues and I take the opportunity to go by helicopter to visit the *Fedorov*. The Russian icebreaker left Tromsø one day after us, so we haven't seen her yet. And communication between the ships isn't easy—the distance has been too great for direct radio contact, and the satellite link crackles and breaks up. But it's important that we discuss with our Russian colleagues our plan for advancing into the ice.

As we'll be flying over open water, we put on our orange survival suits; if the helicopter has to make an emergency water landing, they'll increase our chances of coming out alive. They are air- and watertight. I kneel down to expel excess air, just like I do before every flight. After that, the suit fits like vacuum packaging. We put on our helmets and board one of our two BK117 helicopters, ready and waiting in the *Polarstern* hangar.

The helicopter takes off gently and tilts to the side as it turns toward the *Fedorov*. It's noisy inside the helicopter, so we communicate via microphones and headphones built into our helmets. The broken ice lies on the dark ocean like a mosaic. After around ten minutes, the imposing red-and-white hull of the *Fedorov* appears beneath us.

The *Polarstern* and *Fedorov* are both around forty years old. Despite their age, they are two of the best research icebreakers in the world. I meet my colleagues in the conference room, among them Thomas

Krumpen from the AWI, to whom I have entrusted the management of the *Fedorov*, and Vladimir Sokolov. In his mid-sixties, Vladimir leads the High-Latitude Arctic Expedition department at the Arctic and Antarctic Research Institute (AARI) in Saint Petersburg. He has the greatest possible expertise in polar logistics, based on decades of experience working in the ice. I have known Vladimir for a long time and value him—and his advice—more than anything else. Now we need to discuss where and how we are going to find an ice floe for the long drift—the ice floe that will be the expedition's home for the next year.

Our experiences so far and satellite data analysis suggest it will be tricky to find a sufficiently stable ice floe. We will need to search a huge area, and I have a plan: We will split up to cover the widest area possible. The *Akademik Fedorov* will head for the region at around 85° north, 120° east. From there, they can use their Mi-8 helicopter (which has a greater range than our BK117s) to visit and sample several potential ice floes in that area. We will take the *Polarstern* to 85° north, 135° east and take a closer look at the ice floes there.

September 28, 2019: Day 9

IN THE EARLY MORNING we retrieve the fourth instrument from the sea. From now on we are heading north, straight to the point where we aim to be confined in the ice. After crossing a few areas of open water, we soon find ourselves moving through solid sea ice. We have reached the central Arctic ice cap, our habitat for the next year.

I spend most of the day and night studying the latest satellite data, which shows the ice conditions at our destination. The images come from radar satellites, which transmit radar waves and measure the backscatter from the surface of the ice. They clearly show the ice structure, channels of open water, and the outline of intact ice floes.

In early fall, Arctic ice is made up of individual "ice islands" that have survived the summer. Between them are large areas of ice debris, ice masses that have been worn down by constant movement and shifting to become an amalgamation of small, intact ice floes, chunks of ice, and slush. We need an ice island to set up our research camp—a large, intact floe that can hold our infrastructure, provide a firm position for the ship, and offer a certain degree of stability all year round. It should be at least three feet thick, but more would be better. It should be surrounded by areas of thin ice and sections of open water in which new ice is forming; we are interested in all forms of ice.

The satellite images can only locate the ice islands; to find out how thick they are, we need to pull our GEM across the ice on a sled or fly the EM-Bird over it in a helicopter. The GEM and EM-Bird are electro-magnetic sensors that can measure the total thickness of the ice. But there's only one way to see the inner structure of the ice and estimate its stability and load-bearing capacity: access the floe and take ice core samples.

Our colleagues on the *Akademik Fedorov* have begun to do just that. They have already reached their search area because they didn't have to retrieve any instruments from the seabed. As their initial reports come in, our fears are confirmed.

We know that, once again, the Arctic has had an extremely warm summer. And now we bear witness to the devastating consequences. The first floes examined are only about twenty-five to thirty inches

▶ The *Polarstern* on course for the central Arctic ice cap.

thick on the whole, and only the upper halves offer a degree of stability. The warm ocean water has melted and eroded the ice from below, and now it is riddled with holes and meltwater channels, like a sponge. The lower half of the ice is full of holes and barely attached to the firmer ice above it; it adds nothing to the ice's stability.

Plus, the floes consist largely of melt ponds that formed in the summer, melted all the way through, and are now freshly frozen over with eight to twelve inches of new ice. At the end of the summer, these floes have more holes than Swiss cheese. We might not be able to see the holes now that fall has started—because they've just frozen over—but the surface isn't stable. We can't attach ourselves to a floe like this. It could crumble beneath our feet at any time, and the first storm we encounter would push the ship right through it.

We haven't looked at many floes yet, but I have serious concerns. How are we supposed to complete this expedition if all the floes are like this? It's a definite possibility; why should they be much different? After all, they've spent the same summer in the same sea region and look pretty similar on the satellite images. Is our expedition doomed before it's even begun?

I soon realize that what we need is a special ice floe. There's probably not much point in examining dozens of identical floes just to rule them all out. We need that one, exceptional piece of ice that will allow us to complete our expedition, that offers stability, that can act as a base to expand our work over the winter as the ice becomes more stable overall, and to which we can retreat next summer when the thin ice melts and breaks. We need to find an ice floe as unique as a snowflake, that sets itself apart from the rest. We need it to save the expedition.

I spend hours poring over the satellite images, trying to understand what I'm seeing. The ice floes are dark, rather uniform islands in the otherwise lighter ice. Shades of lighter gray indicate the rugged surfaces of the ice debris between the floes; these surfaces scatter back the radar radiation more efficiently. But out of the vast Arctic, one little floe piques my interest. It appears on one of the dozen satellite images.

It measures around two by one and a half miles and looks mostly dark, like all the ice floes. But there's a large core in its north section measuring around half a mile by a mile. On the satellite images, this core is as light as the ice debris around the intact floes. But why would an otherwise intact floe have a lake of rough ice debris at its center? How could this have developed without the floe being destroyed around it? On closer inspection, it's actually lighter than the typical broken ice in the spaces between the floes. As I study the images, I keep coming back to this particular floe. Is this the snowflake I've been looking for?

Late at night, I decide that we should sail to this ice floe. I present it to the expedition members at our group meeting the following evening. Along the way I want to look at a series of typical ice islands in this area and chart a zigzag course from floe to floe that will ultimately lead us to the one with the light-colored core.

September 29, 2019: Day 10

AS MORNING DAWNS, we approach our first potential floe at 84° north, 129° east. But where exactly is it? Our data is from the satellite's last pass, and that was almost twenty-four hours ago. Since then we have been working blind; the ice continues to drift at varying speeds, often moving erratically with the tidal cycles and natural oscillations of the Arctic ice cap. On the bridge we gaze intently at the ship's ice radar, which maps our surroundings, and compare the structures it shows with those on the last satellite images. I gauge the probable drift and identify smaller floes on our approach. After a little practice, we become pretty good at matching the floes on the satellite images to the green specks on the ice radar. Slowly and cautiously, we approach the floe until it's clearly shown on the radar.

While the ship decelerates, we stand on the bridge and look for the floe. Eventually we can see it without the radar: a large, almost smooth surface before us in the endless icy wilderness. The ship stops—we don't want to destroy a potential candidate—and the helicopter is readied. I take off with a team of ice specialists and a polar bear guard.

As we fly toward the floe, I tell the pilot to take a wide left turn around it so that I can inspect it closely from the copilot's seat. The floe is huge, permeated by a pressure ridge running east to west and criss-crossed by a few smaller ridges. It also has a channel of open water on its eastern side. That would be a great spot for our research camp; the older ice on one side of the ship, and on the other side the new ice that is currently forming on the channel. In fact, it's perfect!

But is the floe thick enough for us? I decide to land in three places to measure the ice thickness. The pilot carefully touches down on the first landing spot. Nobody knows how thick the ice is or whether it will hold. The pilot slowly reduces the lift, ready to take off immediately if the ice should break. Gradually he sets all the helicopter's weight on the sea ice—it holds.

We get out. We are actually standing on the ice cap at the northern end of our planet. Just above the horizon, the sun shines through thin clouds, bathing everything in yellow light. It's −8°C (about 18°F) with only a slight breeze. The ice at our landing site is remarkably flat, and the wind gently blows a few snow crystals around our feet. The *Polarstern* is a dot in the distance: our home, safe and warm.

▶ A FORTRESS ON THE OCEAN

Arctic sea ice is constantly observed by satellites from various nations. They measure radiation and send radar waves to Earth. The conditions on the Earth's surface are determined based on how it returns these waves. For example, individual floes can often be identified by their contours, as they are on satellite images. Researchers don't yet know all the reflective properties of ice, so these images are sometimes puzzling, just like the ice floe's light core; until the team see it with their own eyes and measure it for themselves, all they can do is speculate. The measurements taken during MOSAiC will contribute greatly to understanding satellite images.

After a brief pause to enjoy the moment, we start drilling our first hole. Seconds later the drill is through. It's not looking great; there's hardly any ice beneath our feet! We lower the lead into the hole and pull it up so that the little weight wedges beneath the ice, allowing us to read the thickness off the tape measure—not even sixteen inches! We quickly drill a few more holes and take measurements around our landing site. They all return similar values. I push away a bit of snow with my boot. The surface of the ice shines wetly. I take off my glove, run a finger over the ice, and do a quick taste test. It's salt water. I repeat this a few times and find wet, salty ice every time. It's moistened by seawater throughout—the ice is thin, and salty seawater trickles to the surface from its many little cracks.

▶ Exploring the first ice floe with the helicopter.

The other two landing sites are the same, with similar thicknesses and a consistent upper layer of wet ice. It's immediately clear that we won't be spending the winter here!

We fly back to the ship. The mosaic of ice shimmers in pink as the sun sets. Older ice ridges and the outlines of ice floes create an endless tangle of ribbons all the way to the horizon. New ice is forming in the darker areas between them.

Because the ice is constantly moving, the sea ice keeps on tearing, creating areas of open water. Now that fall is starting, the water is covered by new ice. First, individual ice crystals grow in the water. They float and turn the thin layer of ice on the surface into a thick blend of water and ice crystals. This "slush ice" dampens the ripples on the water's surface, the little waves that usually play and dance in the wind. The ice crystals make the water sluggish; it sloshes dully between the ice floes as though covered in a thick layer of oil.

Now the ice crystals cake and the water between them begins to freeze. When it's windy, this freezing slush ice is repeatedly broken into small pieces that collide, curling their edges. This creates a wild jungle of little ice floes that look like pancakes, hence their name—"pancake ice."

If there is less wind and fewer waves, the individual ice crystals form an initial, fragile layer of ice that is dark and transparent at first; this is known as "dark nilas." More ice crystals gradually attach, making it lighter and more stable and turning it to "light nilas." This layer becomes more durable over time; once it's about six inches thick, you can walk on it, depending on the conditions.

But these layers aren't all that stable. If areas of dark nilas or light nilas are subjected to lateral pressure, they simply slide on top of each other. A characteristic pattern—"finger rafting"—can often be found in vast areas of new ice, and it really does look like intertwined fingers.

As we fly back to the ship, these various stages of new ice spread out below us. Between the older, consolidated ice floes, we can see pancake ice and large areas of dark and light nilas, covered in beautiful finger rafting patterns.

▶ SEA ICE HAS MANY FACES

Sea ice is different from land ice. It doesn't have a stable subsurface. Currents and wind drive it across the Arctic Ocean, pull it apart, and push it back together. It first appears at around −1.5°C to −1.7°C (around 29°F); the salt in the water lowers the freezing point. When it freezes, channels of highly concentrated salt water form within it. Sea ice is usually milky rather than clear; highly specialized organisms live in its many cavities, fissures, and saltwater channels, and algae carpets the underside of the ice. Sea ice doesn't simply solidify as it freezes; the process is complex and depends on a medley of conditions, including wind, movement, and temperature. This is why the ice takes on so many forms during this early phase, including slush ice, pancake ice, and light nilas.

▶ Top left: Water surface covered in slush ice.

▶ Top right: Pancake ice.

▶ Bottom left: Finger rafting pattern in light nilas.

In the endless icy mosaic, the *Polarstern* shines in the distance, illuminated by the low-lying sun and drawing ever closer. We circle our little world and land safely on the helideck.

Back on the ship, we receive news from the *Akademik Fedorov*. They have examined several more floes. On the whole, their findings are as devastating as ours. The floes are all the same—too thin and too unstable.

In the evening I sit down with our remote sensing experts, the world's leading minds on interpreting satellite data from sea ice. Now that we have seen the floes, we can read their images better. All the floes here are thin and moistened with seawater. Their wet surfaces absorb the satellite radar signals, rather than backscattering, which is why the floes all appear on the radar images in the same shade of dark gray.

It's so disheartening. We have to assume that all these floes are as thin as the one we examined. None will be suitable for our expedition. What are we supposed to do?

At the very least, the theory put forward by our remote sensing experts supports my idea about the light area on the special ice floe. Could this really be a thicker area in which ice compression has repeatedly pushed pieces of ice on top of one another, piling them up? Does its rugged surface backscatter the radar signals better? Is its surface drier, and is this why it shows up lighter on the satellite images?

I put all my eggs in one basket. It would be a waste of time to explore other ice floes around here. I call off our previous zigzag route. Instead of spending the night here and examining more potential floes in the morning, we head directly for the special floe. It's our last hope. We travel through the night.

What's Happening to the Arctic?

IN MARCH 2008, I returned to the Arctic after several years of research in the tropics. I had traveled in the Western Pacific, set up a new research station on Palau, and joined and led expeditions to Borneo and Nepal.

Now the Kongs Fjord lay below me, on the west coast of Spitsbergen, 80° north and just six hundred miles from the North Pole. We were flying to the Arctic research station, which has come to feel like home over the years. I thought about the first time I traveled to the station, in an equally small aircraft, in 1992. Since then, I had come to know the landscape like the back of my hand.

In my mind I skied across the frozen fjord to the huts on the other side, as I did so often in the 1990s. Then, in 2008, I cast my gaze over the fjord—and I was shocked. What was going on? It was March, winter in the Arctic; there should be nothing but snow and ice! That's how it had always been before. But now, waves undulated on the open water, glittering in the sunlight. Clearly, I wouldn't be skiing across the fjord.

After landing, I went and stood on the shore. I felt anxious, unable to escape the sense of foreboding—this was a world on the brink of extinction. In the past, this landscape would have spent the winter frozen solid, with blue ice and white snow as far as the eye could see. Now open water rippled at my feet. In the past, the boats would have been mothballed when fall came. Now, in the middle of winter, they rocked gently on the water. In previous decades, the fjord froze every winter; you could ski to the deep-blue icebergs that had calved off the glaciers the previous summer, drifted into the fjord, and frozen in the ice. The icebergs were still there—but now you needed a boat to reach them.

And what about the glaciers? I looked for the edge of the glacier but couldn't find it. Back in the station, I compared the photos I had just taken on the flight with similar pictures from earlier times. It became clear that the ice masses were retreating faster and faster. Today, the glacier edge is over a mile farther inland than on my first photo from 1992.

This was confirmed by my first skiing excursions on land, which led me into an unfamiliar landscape. Where previously I would have climbed the Brøggerbreen—the glacier closest to the station—instead I saw the moraine landscape left behind as it retreated, its little hills, ridges, and piles of rocks extending as far as the eye could see. The glacier was gone. I found the tongue much farther inland.

▶ View of an iceberg frozen into the winter sea ice on the Kongs Fjord in March 1992. In the 1990s, the fjord froze solid every winter, and people rode across it on skis or snowmobiles.

▶ The same spot in April 2018. The fjord hasn't frozen for a decade. Where previously winter would have been the preserve of ice and snow, now the open water of the fjord ripples all year round. Its icebergs melt faster, and ski tours have been replaced with boat trips.

▶ The tongue of the Kronebreen in April 1996. Off the glacier, the Kongs Fjord is solid ice.

▶ The Kronebreen in April 2008. The glacier front has retreated, and the fjord is covered in just a thin layer of new ice.

▶ The Kronebreen in April 2018. The glacier tongue has now retreated by two kilometers (1.2 miles), and the Kongs Fjord is open and free of ice.

Just one year before, the sea ice had reached a historic low, and we all knew about Arctic warming. But seeing it with my own eyes was something else entirely.

What happened here?

The answer can be found in the research station's data. As a yearly average, this region has grown around 3.5°C (6.3°F) warmer since the mid-1990s—much faster than the rest of the planet—and a staggering 7°C (12.6°F) warmer in winter. In the mid-1990s, average annual temperatures were around −5°C (23°F); today, they are near freezing. It's conceivable that they might soon be above freezing. Positive average temperatures in the middle of the high Arctic!

As a whole, the Arctic is actually warming more than twice as fast as the rest of the world; winter is warming faster than summer and Spitsbergen more than anywhere else.

And this means that the sea ice is also retreating. Based on the minimum sea ice at the end of summer (September), it has retreated by more than 40 percent in the last forty years. If we compare current data with ice-thickness measurements taken by submarines in the 1960s and 1970s, we can see the ice is around half as thick as it used to be. It has already dwindled to little more than a quarter of its previous volume.

The remaining ice breaks faster and melts earlier. This means there is now much less old sea ice, which usually spends years drifting across the ocean, repeatedly deforming, and being pushed together to form thick ice floes. Today, almost 90 percent of Arctic ice is young ice (no more than two years old). In the mid-1980s, around half of the Arctic was covered in older ice. And the very old ice—four years and older—has practically disappeared; in March, it accounts for around 1 percent of sea ice.

The ice in the north is retreating, growing thinner and more brittle. This is becoming very apparent as we continue our expedition.

September 30, 2019: Day 11

WE REACH OUR DESTINATION at 4:30 AM and stop in thin ice a safe distance from the floe. I use the binoculars, radar image, and satellite data to find a temporary mooring point. We want to spend an entire year observing and studying this floe in its current, untouched state, so we mustn't destroy the ice or crush the snow surface. Captain Stefan Schwarze skillfully maneuvers the *Polarstern* through the ice to our chosen spot without damaging the precious floe. I know Stefan well from earlier voyages on the *Polarstern*. He has been sailing the ship through the Arctic and Antarctic for decades. He will spend the first six months on board; Captain Thomas Wunderlich will take over for the second half of the expedition.

We have chosen a good position. To starboard we have an even surface suitable for unloading equipment for detailed surveys. Before us is the mysterious light-gray area from the radar satellite images—the enigma that brought us here. With the binoculars, I can see mountains of ice blocks and chaotically interwoven ice floes. It looks huge and thick—we've hit the jackpot!

It's definitely worth exploring the floe more closely. We quickly use the starboard crane to get a team on the ice with snowmobiles and equipment, including the GEM (the device that continuously records ice thickness as it's pulled across the ice, for example by a snowmobile). Over the next few hours, the GEM will complete several loops of the floe and map the ice thickness.

Our floe is made up of three sections—flat and thin at the far north, thick compressed ice in the rest of the northern third, and another flat and thin area—a vast one—in the south. In the thin-ice areas, this floe is just like all the others that we and the *Fedorov* have examined. Here, up to 70 percent of the surface is even, with an ice thickness of about twelve inches; these areas are melt ponds that form on the ice in the warmth of summer, then melt through completely, and have now frozen again in the fall. Between them are strips around thirty to forty inches thick, their lower halves entirely porous. But the compressed core in the northern third, measuring half a mile by a mile, is a real find! The snowmobile and GEM can't cope with the craggy surface, but where they can negotiate the terrain, the ice tends to be at least ten feet thick, and the high ridges between these flatter areas are sure to be even bigger. We decide to continue exploring in the morning, so we remain by the floe overnight.

▶ Ice formation on the MOSAiC floe during initial explorations.

October 2, 2019: Day 13

YESTERDAY WE TOOK a closer look at the ice floe. We now have suffi-cient data, so last night we set off to meet the *Akademik Fedorov*. We reach our agreed meeting point in the morning and are delighted to see that the ship is already there. We come alongside and our Russian colleagues use their mummy chair to join us. Mummy chairs are used with ship's cranes to heave small groups of people from ship to ship, onto the ice, or onto land. Our mummy chair is a practical orange steel box that I have used many times. But the *Fedorov*'s chair is truly a sight to behold. It looks like a cross between an enormous birdcage and an oversized fish trap. The Russian delegation get into the cage and the crane lifts them over.

After welcoming them with hugs and laughter (how often do you meet old friends so close to the North Pole?), we withdraw to the ship's Blue Salon. We now have to make the most important decision of the whole expedition. The tension is palpable. We are all asking the same question—where should we allow ourselves to be frozen into the ice, and which ice floe suits our purposes? This decision will determine the fate of the expedition for the coming year. It will determine whether we drift on the right course and whether our new icy home will turn out to be sufficiently stable. Years-long studies of drift statistics have been leading up to this moment. I have memorized all precalculated drifts for all possible starting points. I know I would like to start at around 85° north, 135° east, the sweet spot that balances all key parameters. When viewed from Asia, the floe must not drift too far beyond the North Pole; otherwise we will be too far away for the aircraft that are due to bring supplies in early April if no more icebreakers make it through the winter ice. If we are too far on this side of the North Pole, we might end up in Russia's exclusive economic zone (EEZ), for which we do not hold a research permit. If we start too far north or west, the ice might spit us out too early. And if we're too far east, there's a risk of getting sucked into the Beaufort Gyre, the ice vortex in the

Beaufort Sea—ice can drift into it over the years, and extricating ourselves would be tricky.

Although I'm well prepared for this crucial decision, I'd like to hear what my Russian colleagues have to say. Nobody knows more about selecting stable ice floes in the Arctic. Our friends have spent decades setting up little drift camps on the Arctic ice to be carried through the North Pole region. The teams that lived in these little wooden huts were entirely at the mercy of their ice floes. They didn't have a secure ship to retreat to in an emergency.

The drift camps were discontinued just a few years ago. The ice floes of the new Arctic aren't thick or stable enough to risk such a venture.

Our Russian colleagues are now enthusiastic partners in MOSAiC. Through them, we can harness the expertise and experience of all the people to have ever selected those ice floes. We sit together in the Blue Salon and pore over our collective ice floe measurements. As usual, conversations take place via our interpreter.

▶ The MOSAiC ice floe is announced in the Blue Salon on the *Polarstern*.

First, our colleagues from the *Fedorov* present their findings from dozens of floes. It's depressing. Not a single floe is even remotely capable of providing our ship with a stable base and bearing the load of the research camp we plan to construct. The ice floes fall into three basic categories: "won't work," "definitely won't work," and "don't even think about it."

We then present our findings in detail. The others are visibly impressed.

After the presentations, I ask Vladimir Sokolov for his opinion. Vladimir's words are always sparing, deliberate, and 100 percent reliable. He quickly consults with his experts on sea ice physics. Then he summarizes his view of the situation, pausing to let the interpreter speak. He confirms my assessment that practically all the floes we have found are unsuitable for the expedition. He is pleasantly surprised that we have found this one special ice floe, and strongly recommends that we use it for our drift. When he finishes, his ice experts nod unanimously. This tallies with my own views, so I announce that this will be our floe. Typically for the MOSAiC expedition, this momentous occasion meets with no great fanfare. We express our delight with smiles and nods. We've found our home for the coming year!

Vladimir's considered and knowledgeable contributions were invaluable as we prepared for the expedition, and I am more indebted to him with every day that passes.

The moment doesn't last long. It's as though someone has flipped a switch; in the very same minute, we start to plan the next steps in the expedition. We've chosen our ice floe and there's no time to lose. Winter and night are almost upon us.

Now we have to:

1. Set up the unstaffed distributed network (DN) of instruments and measuring devices that will be located up to thirty miles away from our floe. The *Fedorov* team need to establish this network as quickly as possible.
2. Transfer cargo and fuel from the *Fedorov* to the *Polarstern*.
3. Construct the central research camp.

A team is immediately assembled to assess the ice situation around our floe with the aid of high-resolution satellite data. Within two hours, potential locations have been identified for the main DN stations, and the *Fedorov*'s Mi-8 helicopter has gone exploring.

Meanwhile, I fly back to our floe in our BK117 to get a good aerial overview. We now need to decide how and where we will moor on the floe for the winter. I am joined by the team responsible for laser scanners and infrared cameras; we urgently need their measurements to create a complete 3D map of the floe and plan the research camp.

In the meantime, the crews pass over the discharge hoses, which will spend the coming hours pumping hundreds of tons of fuel from the *Fedorov* that the *Polarstern* will need for the long winter ahead.

Our bar, the Zillertal, opens in the evening. Over vodka and beer, both crews celebrate the successful conclusion of a crucial stage—and the beginning of the next.

3

A NEW HOME

October 4, 2019: Day 15

THIS MORNING THE SKY clears for the first time in days. Suddenly we can see the sun shining just above the horizon. What a sublime sight! Just what we need on this big day; after determining our mooring point and discussing it in detail with the captain, we decide that the ship will be rammed into our ice floe tonight.

We spend the day completing our cargo operations. Yesterday we transferred load after load from the *Fedorov* by positioning the two ships stern to stern. The two heavy PistenBullys—snow groomers—were heaved over to the *Polarstern*, a painstaking task that involved maneuvering the *Fedorov*'s crane past the *Polarstern*'s bridge while both ships swayed in the water. At one point, the ice pressed the ships together so hard that the *Polarstern* had to hastily cast off. We changed position several times so that the cranes could put everything in the right place. Now our ship is jam-packed; luckily, we took on hundreds of tons of fuel from the *Fedorov* beforehand, so it's still within the permitted stability range.

I'm now planning the route to the floe. We don't want to destroy any of the smaller floes because they will later serve as bases for our DN instruments, our network of measuring stations. The problem is that our satellite images are now hours old, and the floes have continued to drift. I use paper and pen to calculate the most probable drift, adjust the coordinates, and give the waypoints to the navigator on the bridge, who enters them in the navigation computer.

By 7:00 PM I'm finished and we can set off.

I stay on the bridge and monitor our approach. The ice drifts erratically, and nobody can say whether the floe positions I calculated are still correct. It's pretty dark outside already, just a thin strip of light hovering over the horizon. The bridge is dark as well, allowing us to better see the shapes of the ice and concentrate on the ice radar, which outlines the floes vaguely. On the radar I identify the DN floes past which we need to navigate. The planned route works well; just once, thick ice forces us off course to the north, dangerously close to one of our chosen floes. We move the rudder to starboard, but the ship doesn't want to comply. The *Polarstern* can be pretty headstrong in thick ice. Luckily, a channel opens up that allows us to continue while narrowly avoiding the DN floe.

We sail around the floe allocated for our large research camp—the floe in which we want the ship to freeze—at a safe distance. It would be disastrous if we failed to see it in the darkness and accidentally broke through it! Soon its outline appears on the ice radar and we use the image for orientation. To set up camp successfully, we need to get the position right and bring the ship to stillness. I have also calculated our compass course and the final route for the maneuver. What we see fits with our plans; we are on course to land perfectly in our floe, somewhere before us in the dark.

The captain and I quietly discuss what's happening, making occasional, tiny adjustments. Otherwise, the bridge is silent, even though some colleagues have joined us. Everyone is focused on this crucial moment; we must not destroy the ice when ramming the ship, and we need to sail into the right part of the floe. It will be our home for the next year.

▶ The *Polarstern* is in position in the MOSAiC floe and stable in the ice.

Through the window, a large stretch of ice appears before us in the searchlights. This must be our ice floe!

Everything rests on the captain's vast experience. If you looked up "sea dog" in the dictionary, there'd be a picture of Stefan Schwarze. He knows more than anyone about the ship and how she behaves in ice. And this knowledge is crucial. We need just enough momentum when breaking into the floe. Too much and we'll shatter the ice. Too little and we'll stop too soon. Gingerly, he pushes the control lever. The colossal ship follows his gentle cues and approaches at a carefully judged speed.

Now the floe is right in front of us. I use the binoculars to check the position of the flags we left here during our initial measurements. Our course is perfect. A quick nod to the captain, who metes out the power precisely. The bridge rattles and shudders as we break into the ice floe. As planned, we smash past an AIS station—a transmitter we left here on our first visit—until it's just off our starboard side. A final quick, quiet talk with the captain, and we steer the bow slightly to port, just a little bit more. Stop!

October 4, 2019, 10:47 PM: we are still.

The ship has stopped in the solid ice and appears to be stable. We gain an overview of our position from the windows and outer deck and leave the engines idling. After we stopped, the ship moved back a few yards; the bow had been raised by the ice and then slipped back down. Now the ship is motionless, like a rock. Her propellers are no longer needed to maintain her position. For safety's sake, the captain leaves the engines idling for a few more hours so that they're available if the ship should move. Eventually he switches them off and the ship falls silent; we keep one little auxiliary engine running to provide power and heat.

Then we all retire. This may be a significant evening, but the atmosphere is muted, and we can't help but dwell on what the next few months might bring.

I spend much of the night poring over the laser scanner map of our ice floe. I go to the bridge, using the searchlights to identify ice formations in the darkness—formations that also appear on the map—and triangulate our exact position and orientation in the floe.

October 5, 2019: Day 16

THE NEXT MORNING, we feel like visitors to an alien planet, and not just because of the vast white plain and bizarre ice formations; we are the first humans to set foot on this terrain.

We are in the eastern section of the floe, to the south and on the eastern edge of its core of solid, compacted ice, which we are now calling our "fortress." This part of the fortress has a high pressure ridge—the "outer wall"—that separates it from the flatter and thinner part of the floe on which we are currently standing. A smaller ridge branches off from the outer wall and goes straight to the bow of the ship. This is what stopped the ship in its tracks. As planned, there is now an area of flat ice to starboard between us and the outer wall; this will be an excellent logistics zone for unloading the ship.

▶ Ice sculpture on the MOSAiC floe.

Together with the group leaders, I have developed an initial plan for our research camp. The team leaders may be very different personalities, but they are all passionate about this massive undertaking. Marcel Nicolaus (my co-cruise leader and colleague from the AWI) oversees the Ice team. When we were preparing for MOSAiC, he was instrumental in moving our plans forward and developed many processes and procedures. Allison Fong, also from the AWI, is the Ecosystem team leader, covering everything that lives in the Arctic. This expedition will earn her a new nickname, "Chainsaw Alli": whenever someone needs to make a hole in the ice, she'll be there, wielding a chainsaw. Katarina Abrahamsson, from the University of Gothenburg, is in charge of Biogeochemistry—studying how chemical elements flow through plants, animals, and the environment—for the first phase; Ellen Damm leads the team but won't come aboard until the second phase. It's great to see Katarina again—we traveled the Antarctic together way back in 1994 and have lost track of each other since. But polar research is a small community, and paths always cross again eventually. Ying-Chih Fang (also from the AWI) is representing the small Ocean team, which is managed by Benjamin Rabe (he won't be joining us until later). And then there's Matthew Shupe, known to all as Matt, from the University of Colorado and the US National Oceanic and Atmospheric Administration (NOAA). He is my co-coordinator on the Atmosphere team and the "mayor," as it were, of Met City, one of the sections of the research camp, where he shares responsibility for dozens of research projects. Matt was one of MOSAiC's first visionaries and helped to develop many of our plans.

I assemble a small team to explore the floe in more detail and choose locations for the main research infrastructure.

The gangway is lowered and the route to the ice is clear. We start by walking along the little ice ridge to the outer wall. The core of our fortress stretches out behind it, a vast and rugged icy landscape with linear ice ridges and bizarre, lone ice formations rising up into the sky. We christen one area the "sculpture garden"; it really does look like something from an enchanted castle.

We plan to build the main part of the research camp in one line with its main axis along the fortress. A perfect setup. The power lines and main pieces of infrastructure are situated along the massive ridge of the outer wall. Some of the measuring instruments can branch off this axis, the "spine" of the camp, on the thin ice that dominates the area   after all, we want to study the conditions on typical Arctic ice. The main power line and fiber-optic data cable run from the bow of the *Polarstern* along the smaller ice ridge for around two hundred yards, straight to the outer wall; they then round the "corner" (we all use the English word, even in German) and continue along the spine to the end of the camp. That's how it's supposed to work.

We follow the outer wall on foot. Matt finds a perfect spot for Met City right at the end of the planned camp. He decides where to put the meteorological masts and the Met City research hut, home to the computer that controls the measuring instruments. The hut is on the outer wall, the mast in the area where the wall transitions to flat ice. The Ocean team establishes Ocean City along the main axis between the corner and Met City.

The Remote Sensing Site and its radars won't be far away, on the thin ice near the outer wall in the direction of Met City. I choose a spot for Balloon Town near to Ocean City, allowing the two to share a power-distributor node along the main power line to Met City. Balloon Town doesn't necessarily have to be situated on typical ice; it will operate the tethered balloon that measures the atmosphere hundreds of feet above the ice. I select a flat but still large area of the fortress beyond the outer wall. The large, immobile hangars should be safe here. In fact, Balloon Town will turn out to be the safest spot, and never under direct threat from cracks or new pressure ridges (unlike almost every other part of the camp). As we make our plans, we read the ice, trying to distinguish stable areas from zones with frequent faults, constantly comparing our position with the structures on the laser scanner map of the ice floe. You quickly lose your sense of distance in this icy landscape, which has very few orientation points. Nevertheless, we soon manage to align the landscape we see before us with the ice floe map, improving our understanding of the ice structures. We drill frequent holes in the ice beneath us and check its thickness. If uncertain, we feel our way by shoving heavy metal rods into the ice in front of us to make sure it will take our weight.

Marcel Nicolaus, our Ice team leader, seeks a location for ROV City, where our remotely operated vehicle (underwater robot) will dive beneath the ice. This area must not be disturbed by our other work and should be characteristic of the surrounding ice, so the outer wall won't be suitable. Despite the extra risk, Marcel chooses a spot away from the main axis, on the thin ice around five hundred yards in front of the ship. Sometimes the best science means accepting risks. ROV City will have a separate power line and a dedicated data cable. This is the last of the larger stations—we now know the basic layout of the ice camp.

Having focused totally on our work, we're done for the day, and suddenly the thin layer of cloud breaks apart. It's around midday. The sun appears just above the horizon and bathes everything in pink light. Arctic light in fall—it's breathtaking. We sit in the snow on a ridge of ice and soak it all in. There's hardly any wind, and all we hear is an occasional quiet creaking when the ice moves beneath us. Nobody speaks. Such moments will stay with you forever, and are best enjoyed in silence. After all, there's no way to put them into words.

A City in the Ice

IT'S BEEN 126 YEARS since Fridtjof Nansen set out to freeze his wooden ship into the ice. Without his incredible achievement, we may not have known that such an expedition was possible. Now we are the first to replicate it with a modern research icebreaker.

But the indomitable *Polarstern* is just one part of our project. She is our haven at the center of an entire city constructed on the ice. Many years ago, we wrote our initial research and implementation plans, explaining how we could achieve our scientific goals in these logistical circumstances. Since then, our plans have become increasingly detailed and more partners have joined, bringing with them projects that we need to accommodate on the ice. We had a rough map of the research city before we'd even found our ice floe.

Of course, we now have to align this plan with the local conditions. And it works—our floe has space for every component and offers good prospects for all our projects.

Starting from the ship, a path runs along the small pressure ridge at three o'clock straight to the fortress, then bends sharply and leads more or less directly to the most remote point, Met City. Met City needs to be far away because this is where we measure air currents and other factors that could be disturbed by the ship's hull, depending on how wind is blowing. Met City takes an enormous number of measurements. Dozens of instruments record all relevant atmospheric parameters: the energy flows carried by the smallest of turbulences, solar and heat radiation, and the aerosols and clouds above us. Ocean City is located on the left just around the corner; this is where the CTD rosette—a probe for conductivity, temperature, and depth—is lowered into the ocean to take water samples. Its much larger brother has to be lowered into the water right next to the ship with the *Polarstern* cranes and won't be used until later. With many other instruments as well, Ocean City is a full observatory for all key properties of the ocean. On the outskirts of Ocean City, in the direction of the ship, is the Remote Sensing Site, with scatterometers and radars on the ice similar to those on satellites. To the right of the path is the Balloon Town field, where our tethered balloon (and its large hangar) deliver constant data from the lower thousand feet of the atmosphere. The power lines and data cables run along this path from the ship to Met City, as does the little snowmobile road.

ROV City is straight ahead of the bow. This is the launch site for the underwater robot, which films the landscape below the ice, measures the incoming solar radiation, takes water samples, surveys the topography of the ice's underside, and studies its ecosystem. We have two ROVs—"Beauty" and "Beast"—in case one fails.

As well as these fixed installations on the ice, we have sampling sites where we will take ice and snow samples throughout the year; the sites have already been reserved and must be kept free of traffic. We will also take mobile measurements, including constant laser scanner measurements, to observe changes in the ice and snow topography. The dozens of fields for snow measurements and samples, and the areas in which ice cores will be taken all year round, must remain partially in absolute darkness—the light-sensitive microorganisms in the ice and water will react even to the artificial lights of our ship. For these experiments, we set up a dark sector miles away from the *Polarstern*, in the constant shade of high pressure ridges. All in all, over a hundred complex climate parameters will be recorded continually throughout the year. Many others will be added in specific phases.

We will examine our surroundings in as much detail as possible—the air, the sea, the snow and ice between them, the ecosystem, and the biogeochemistry—allowing us to determine their mutual influences. This is MOSAiC's unparalleled advantage: we will observe the entire Arctic system in a defined zone and get to know the processes at play. Our distributed network extends this zone even farther. Climate models divide the Earth's surface into a grid of three-dimensional cells, and our study area is approximately the same as a single grid cell. This links our measurements to climate modeling—an invaluable resource. Up to now, we have had far too little data about the climate processes of the central Arctic, and no data at all from the winter season because even the strongest research icebreaker can't get that far north. The data we are collecting now will benefit researchers for generations to come.

AFTER A FEW HOURS of exploring, we return to the *Polarstern* feeling satisfied. I decide that as long as the sun is still above the horizon, everyone is free to go onto the ice.

After lunch, we secure an area near the *Polarstern* with a ring of polar bear guards, and then most of us descend onto the ice. Some people simply wander in silence over the unspoiled, snow-dusted surface. Others take photos, chat, or throw themselves in the snow. In the background, the sun glows on the horizon in red and gold—what a blessing on our first day in a new home! It's the last time we'll see the sun. After this, it disappears behind the horizon for good, the last sunset before the long winter.

▶ Their work done, the first-phase team returns to the ship from the MOSAiC floe. The sun sets for the last time and won't rise again for almost six months. The polar night begins its reign.

But we don't know that yet. There's no way of predicting when the sun will make its last appearance at our current position; even if it's entirely beyond the horizon, mirages may still allow us to see it. It's a curious thing—what you see on the Arctic and Antarctic horizon above the extremely cold snow is often deceptive, and isn't always where it appears to be.

I have often been in the Antarctic, surrounded by shelf ice far from the ocean, and have seen beautiful bays with floating icebergs that were anything but real. Mirages trick the eyes into thinking that things far beyond the horizon are actually in places they couldn't possibly be. (See plate 2 of photo insert.)

The same happens with the sun. Sometimes it is reflected in the sky, totally distorted, long after it has set. Even Nansen was confused, and found himself wondering how the sun could still be visible on days when—according to his calculations—it should have disappeared beyond the horizon for the long winter. His report contains a sketch of a square sun on the horizon. He was unable to explain the phenomenon, but it's clear that what he saw was a repeated reflection of the sun on various atmospheric layers, giving the impression of a square.

These striking and bewildering images are caused by air layers of various densities in the lowest part of the atmosphere. There is a very cold layer of air above the chilly ice in the Arctic and Antarctic, and the temperature rises in the lowermost one hundred meters (about 300 feet). This layering is very stable and suppresses any turbulence. Layers can have different temperatures and densities; they don't mix and, if you look at them evenly from below, they reflect the light just like a water surface viewed obliquely from below. In a sense, we are standing in a dense lake of cold air—the cold air on the ground— and looking obliquely into the warmer, less dense air above us, which creates reflections like those on water. These layers will often overlap, leading to bizarre shapes and unreal images on the horizon.

Once the sun has disappeared from view, it spirals ever deeper in its eternal course around the horizon; from the second half of October, even the last little bit of discernible twilight will vanish at midday. As of October 22, it is always more than six degrees below the horizon; civil twilight is over, and the central Arctic will spend many months in icy, black night, hostile to all human life.

We don't expect to see the sun again until March. Since we don't know where we will have drifted by then, we also don't know when it will reappear—the farther north we are, the later it will be. But it'll be March 21 at the latest, when the sun finally rises at the North Pole. What will happen during the long polar night?

▶ BETWEEN DAY AND NIGHT

The Arctic is famous for its polar day and night—summer, when the sun is always in the sky, and winter, a time of never-ending night. They are interspersed by spring and fall, short periods in which the sun rises and sets like it does in the rest of the world. The farther north you go, the shorter these periods are. At the North Pole, polar day blends seamlessly into polar night. Although the sun is below the horizon when polar night begins, it isn't pitch-black from the outset. At first the sunlight still spreads across the sky, creating various phases of twilight. Distinctions are made between these periods according to the sun's depth. "Civil twilight" begins with the sunset and ends when the center of the solar disk is six degrees below the horizon. Artificial light isn't required during this phase. This transitions into "nautical twilight," which lasts until the sun is twelve degrees below the horizon. Stars and constellations are now easily identifiable. This is followed by "astronomical twilight," during which the sun sinks to eighteen degrees below the horizon. From this point on, there are no noticeable light beams in the sky; it can't get any darker. The absolute darkness of polar night can only be experienced from the eighty-fourth parallel and farther north.

October 6, 2019: Day 17

HAVING ALREADY CHOSEN our ice floe, we are ahead of schedule, and yet time is running out. We have just two more weeks of residual daylight; after that, setting up camp becomes decidedly tricky. We need to establish the main infrastructure while we have some light, can maintain an overview of the ice, and don't have to restrict helicopter use—total darkness will prevent us from flying with external loads or landing on unmarked areas of the ice, for example. Plus, the winter is really making its presence felt. The ice is getting thicker and it's high time the *Akademik Fedorov* was leaving, or it might get trapped in the ice too. We give ourselves a week to establish the station network using the *Fedorov*. But first, we need to swap around two dozen expedition members between the *Fedorov* and *Polarstern*, plus a significant amount of equipment and instruments. The question is, how?

My original plan was for the *Fedorov* to moor on our floe as well, a few hundred yards away, so that we could exchange people and materials across the ice. But now we are confronted with the reality of the eroded ice around us.

However the ship approaches, she risks breaking our thin and unstable part of the ice floe. The thicker, more robust part—the fortress—is embedded in the thinner ice and can't be reached without damaging the rest. Even our maneuver to get close to the fortress required us to break some way into the ice floe, inevitably damaging our approach route and risking further damage in front of the ship. If the *Fedorov* moored at our fortress, more damage would be unavoidable, and when she departs, even more of the ice surface would be destroyed. I don't want to take the risk.

So I change the plans again. The *Fedorov* will stay a couple of miles away from our ice floe so as not to disrupt the sensitive environment. Instead, I want to find somewhere safe for the heavy Russian Mi-8 helicopter to land on the floe's thick core, somewhere we can safely reach on snowmobiles from the *Polarstern*. We can then use this helipad on the ice as a hub for our logistical operations and all exchanges. The Mi-8 is far too big and heavy to land on the *Polarstern*'s helideck.

Marcel and I start looking for a suitable landing site. The helipad must have stable ice of at least thirty inches, even and without ridges, and it must be easy to reach by snowmobile from the *Polarstern* while pulling heavily laden Nansen sleds. It's clear that the site will have to be within the fortress. It's the only place with sufficiently stable ice.

Once again, the laser scanner map proves invaluable. But how do you navigate on a piece of ice that's constantly drifting? Absolute coordinates—like those found on land maps—lose all meaning in a matter of hours; the drift can exceed a mile in three hours. After just a short time, a fixed point on the Earth's surface is in a totally different part of our ice floe—coordinates are useless and GPS navigation doesn't work.

We need to think again. In the end, we develop our own coordinate system to navigate the floe and find our way around outside. Navigation must be relative to a fixed point and direction on the ice. We choose the bow of the *Polarstern* as the fixed point and its axis as the direction. Coordinates will then be specified in angular degrees based on the distance from the ship and the direction from the *Polarstern* axis. To keep things simple, we often name directions after the times on a clock face, with the bow representing twelve o'clock. For example, "1,250 meters at 3 o'clock" would be equivalent to starboard abeam 1,250 meters (about 4,000 feet) from the ship. This polar coordinate system helps enormously with navigation and is added to our maps of the ice floe. As long as we have visual contact with the *Polarstern*, we can use the laser measuring feature in our binoculars to determine our distance from the ship and our direction relative to the ship's axis. We can then quickly find our position on the map, note it down, and find it again on our drifting ice floe.

If we are farther away or need more precision, we can use the bridge's navigation systems. On longer excursions, we always take a transponder that shows our current position on the ship's navigation computer. The bridge can use the radio to provide us with the

Polarstern's current position and our exact distance and bearing, allowing us to locate ourselves using the ice map coordinate system.

We use this to follow our route across the ice on our laser scanner floe map. First, we go into the fortress, deep into the area of significant ice pressure that has created this unique piece of thick, solid ice. We pass all sorts of ice formations, some adorned with little icicles as though decorated just for us. One looks like a huge mushroom, another like the teeth of a mighty Arctic monster frozen in the ice. It all seems so surreal that sometimes it's easy to believe the Arctic might be hiding as-yet-undiscovered mythical creatures.

Flat plains extend between the ice formations. On the endless Arctic ice cap, far from the *Polarstern*, the landscape makes you realize just how small you are. You can't help but respect its vastness, aware that there is nothing else for hundreds of miles around you.

This time we are riding snowmobiles, steering through the rugged icy landscape. Every bend reveals new scenery, with ice sculptures never before seen by human eyes; this sculptor doesn't need an audience, or applause.

When we turn off the snowmobiles, a breathtaking stillness descends. We often find ourselves holding our breath, keeping still so that our polar gear doesn't rustle and we can be enveloped by the silence. This is when you become aware of the Arctic's subtle beauty. The low rush of the ice crystals wafting across the snow; an almost imperceptible creaking as the ice moves. This is why I love the Arctic.

October 8, 2019: Day 19

THE WIND FRESHENS as the day goes on, buffeting the waymarker flags we have placed on the ice. It blows snowflakes into our faces that harden to ice crystals in the cold. It burns, like pinpricks on the skin. By midday we've reached level 7 on the Beaufort scale and a severe snowdrift sets in; between the flurries of snow and gusts of wind, we can barely see fifty yards ahead. This would be great camouflage for polar bears. For safety's sake I stop all work on the ice and return to the ship myself. The wind howls around the ship and rattles every structure. We watch through the panoramic windows on the *Polarstern*'s bridge, our safe and cozy haven, a bastion in the forces of nature. But the storm leaves its mark. In the afternoon, when things have calmed a little, a crack appears in the ice, starting from our bow. We watch as it quickly opens, widening and zigzagging through the ice floe as far as we can currently see. Where previously there was solid sea ice, now there is a fissure several feet wide through our surroundings. The ice finally calms down about an hour later. But the crack remains.

▶ The first crack in the MOSAiC floe. Many more will follow as the expedition progresses.

October 9, 2019: Day 20

THE NEW DAY greets us with calm, friendly weather and a good view. We can now see that the entire floe is cracked, a wide arc forming from our bow to the portside. It won't disturb us over there—that's the thin part of the floe—and it's still some way from ROV City; our observatory on the ice doesn't have any infrastructure there. We were wise to set up most of our equipment along the more substantial outer wall on our starboard side.

▶ A polar bear and her cub emerge from the dark polar night to examine the research camp.

With a clear sky and little air movement, the ground quickly cools and the temperature drops to −15°C (5°F). The sun is below the horizon and casts a yellow reflection on the open water in the crack. And the water's smoking! Tendrils of fog rise off it, dancing in the fleeting air current and playing with the light.

This fog is the result of huge temperature differences between the cold air and the ocean, which is relatively warm at just −1.5°C (29°F). The air rises above the warm water as though from the bottom of a cooking pot, and convection cells are formed—little air current fields in which warmth rises. Water vapor rises out of the ocean and into the air, condenses in the cold, and forms wisps of fog. This phenomenon is known as sea smoke; it can be found on bodies of open water on cold and windless winter days. Today's display doesn't last long; soon there is a thin layer of ice on the crack that cuts off the water vapor. We hurry to take samples of the seawater and young ice.

The construction work on the ice is progressing well. By the evening we have dragged our main power line as far as the corner, which will form part of our road across the floe.

▶ After exploring the camp, they play exuberantly.

October 10, 2019: Day 21

IT'S EVENING, DARK already, and the teams have all returned from the ice. Suddenly two bright dots appear on our thermal imaging camera screens. There's something warm out there, and it's moving toward our camp. The ship's searchlights quickly reveal a mother polar bear and a cub less than one year old. They've spotted us, and now they want to look at this strange new thing on the ice. They head straight for Ocean City. The mother still has very light fur, and is probably relatively young herself. This might be her first cub. But she knows how to provide for herself and her offspring; they are both well fed and their dense fur isn't the only thing that makes them fat and round. It's wonderful to watch these majestic creatures.

Ocean City is still under construction, but a hole has been sawed in the ice so that instruments can be lowered into the ocean. The two bears are clearly interested. What sort of hole is this? The breathing hole of a tasty seal? It doesn't smell like seal. Plus it's square and far too large to be a seal hole. And there are toys scattered around. The bears have never seen anything like it.

Ultimately, the toys—our installations—prove more interesting than the hole. The bears explore everything, pull on the flagpoles to stand upright, knock them over, and test pieces of equipment with their teeth. Anything red or orange is particularly fascinating. Could it be animal flesh? A tentative bite into our thick orange power lines proves disappointing; they're rubbery and don't taste good at all. Their nibbles don't damage the cables, which are not yet live.

Polar bears are incredibly curious and playful. Very little scares them—and why should it, when they have no natural predators? Their curiosity helps them to exploit all conceivable food sources in the hostile Arctic. But sometimes they simply appear to be having fun.

I once saw a bear on Spitsbergen, on the frozen Kongs Fjord. He scaled a little iceberg that was frozen into the fjord, sat down on its highest point, and spent a few minutes enjoying the panoramic view of the fascinating landscape—at least that would be the human interpretation. In fact, he was probably sniffing in all directions, searching for the scent of a seal. Then he sat on his hindquarters, slid down a snowy ramp on the side of the iceberg, climbed back up, and did it all again. He proceeded to do this for at least fifteen minutes. I can only think of one reason for this behavior—he was enjoying himself! A polar bear researcher might have another, more rational explanation, but I don't. Moments like this make you feel very close to the animals, and it's easy to humanize them.

Polar bears weigh about half a ton and their bodies are around ten feet long. And yet they move as slinkily as cats. They spring nimbly over cracks in the ice, climb over pressure ridges in a few quick steps, slide effortlessly in and out of water channels, and swim rapidly with powerful strokes. You can tell that they rule the Arctic, that this is their domain. They compensate for their enormous mass with even greater muscular strength, and they move elegantly and at a surprising speed. How unwieldly we are in comparison, lumbering through this environment—a place in which we don't actually belong—in our bulky polar gear that turns every water channel and pressure ridge into an almost unassailable obstacle!

The bears follow the power line toward the ship, finding it all rather fascinating. They enjoy knocking things over just to see what happens. The young bear runs back and forth excitedly, anxious not to miss a thing. They come so close that we can hear him from on board; his noises are incessant, half yelping, half barking, like an excitable puppy, just ten times bigger. He keeps returning to his mother for brief moments of physical contact, as though reassuring himself that everything is fine, although he doesn't seem too perturbed by the situation.

The bears reach the large power distributor node right next to the ship's bow. Its power lines aren't live yet, so there's no immediate danger to the bears.

But the time has come to scare them away. If they get used to our presence and their interest in us outweighs their migratory instinct, they might keep coming back—and sooner or later, this could put both us and the bears at risk. Plus, they won't find any food here. They need to follow their natural lifestyle and roam the Arctic on the relentless hunt for its smattering of seals. Anything else could put them in danger. Experiences from settlements in polar bear habitats show that early and emphatic deterrence can reduce the number of incidents between humans and animals, making things safer for the polar bears as well. If a bear gets too close, if the danger is overwhelming and an attack is imminent, then we may ultimately have to defend ourselves with guns. And we want to avoid that at all costs.

And yet I hesitate for a moment. We could let off flares, a type of ammunition that flashes and detonates loudly. The bears would be scared and run a few hundred yards, but they'd come back. We still haven't explored much of the ice, it's dark, and the area is streaked with channels, which means we can't use snowmobiles to drive the bears away for good. They might learn that although the flashing and banging of the flares is scary, it won't actually harm them—this would reduce our chances of scaring them off permanently.

After a few minutes, however, it becomes obvious that the bears are getting more familiar with our installations. This is the exactly the kind of habituation we don't want to see. I speak briefly with Audun Tholfsen, an expedition member who helps to manage polar bears on Spitsbergen. He confirms that it's time to act. We prepare the flare guns and position ourselves on the bow of the ship, the best place to fire. But the bears suddenly lose interest in the power distributor and cables. They round the bow and pad to our portside, where there is no equipment. They rest, and the young bear snuggles close to his mother after all the excitement.

Unfortunately, this doesn't last long. After exploring our portside, the bears return around the bow to the power distributor. Maybe there's something edible they didn't spot before? The mother starts to shred the plastic covers of the cable splices with her teeth. She might swallow bits of plastic, which wouldn't be good for her. We have to stop them. Immediately I fire the first flare from the bow, aiming between the ship and the bears. The cartridge explodes in the air with a bright flash and loud bang. The bears immediately flee the ship, just as we hoped. As arranged, Audun fires a second flare while I reload and then fire again. We continue to alternate between loading and firing and the cartridges detonate in the air, following the bears at a safe distance as they escape across the ice.

▶ The last shimmer of daylight on the horizon. The twilight phase will end in a few days, and the deep-blue phase will begin. A barely perceptible blue light will continue to illuminate the area before the total blackness of night takes hold.

To make sure they don't get used to this, we have to scare them off for good. We fire eight times, until the bears are out of range of the flares. They stop running but have had enough of us for the moment. They trot away leisurely into the darkness. But I suspect we'll have to do it all again before it has a lasting effect. My suspicions are correct.

October 11, 2019: Day 22

THE DAY BEGINS with a surprise—the crack is gone! The floe has pushed back together during the night, and where yesterday there was a channel covered with a thin layer of new ice, now there is an ice ridge. Ice pressure has forced the thin new ice out of the crack to form a line of thin floes chaotically piled on top of one another. This won't be the only time we see a small pressure ridge forming, a prelude to the huge ice mountains that will emerge.

To my relief, there are no polar bears to be seen. It promises to be a productive day. But after the events of the night, I want to check our surroundings one more time. We prepare the helicopter for a reconnaissance flight.

Just as the helicopter takes off, our two friends are spotted a few hundred yards away on our portside. The mother bear displays totally normal hunting behavior and isn't interested in us anymore.

I'd much prefer to let the bears hunt undisturbed. But if they learn that our environment is safe, the subsequent risk will be simply too great. So we divert the helicopter toward the bears. Polar bears generally shy away from helicopters. Slowly we drive them away from the ship, bit by bit, out of our scent stream so that they won't be enticed by the ship's smell later on.

Then we resume our work on the ice.

October 12, 2019: Day 23

I'VE HAD A restless night. I'm worried about the polar bears.

We closely examine our surroundings from the bridge; there's no sign of them. We lower the gangway and the teams begin their work on the ice. It's snowing a little, but otherwise the conditions are good and the view satisfactory. This appears to be a good day.

But appearances can be deceptive. Around midday, the polar bear guards on the bridge spot two bears just over a mile away on our portside. Clearly, our old friends have returned. Like yesterday, they are hunting for seals; the mother sniffs along cracks in the ice, which seals use to breathe. Occasionally a spot piques her interest. She spends over an hour at one ice hole, not moving a muscle, staring intently at the water, her position never changing. You'd be forgiven for thinking she were stuffed. The cub watches his mother hunting and capers in the snow. Every now and again the bears look at us curiously, without coming closer. But that's about to change.

As we are currently working on our starboard side and the bears show no signs of coming closer, I allow work to continue—but we monitor the bears the whole time.

This is a good opportunity to study their heat signature on our infrared camera systems. Soon it will be pitch-black, and then the thermal imaging cameras will be our only way of locating polar bears in the darkness. We have three camera systems on board. The FIRSTnavy, a state-of-the-art infrared camera developed for the military, sits in our crow's nest, the highest point on the ship. It rotates rapidly and provides live 360-degree images that we view on two high-definition screens side by side. The slight differences in surface temperature represent the ice structures and are reproduced by the camera in varying shades of gray. The warmer the surface compared with the environment, the lighter it appears on our screens.

Sometimes, white zigzags on the screen tell us that cracks are opening in the ice before they are visible—human eyes can't see the warmth of the ocean flowing through the crack and heating the snow. Even in complete darkness, we can closely follow the movements of anyone on the ice.

From a distance, the polar bears start off as bright dots. Holes in the ice also show up as almost-white spots—the water is at least as warm as the outside of a bear's thick fur—so this can be deceptive. We mark bright dots on the screen with little stickers; if the dot moves, it's time to raise the alarm.

Up close, one spot on a polar bear's body glows brighter than the rest—its nose, which is warmer than its insulating fur and shows up well on the camera, but only if the bear is very close.

This system is invaluable. The pin-sharp images are the best that modern technology has to offer. Unfortunately, the system is also highly complex and sensitive. We will soon come to realize just how much the equipment is suffering in the harsh winter conditions.

We also have two more thermal imaging cameras, one at the stern and one that we can swivel and zoom, allowing us to target suspicious areas and follow polar bears once they move out of range of our searchlights and binoculars. These cameras are not as sharp and don't provide panoramic images, but they serve us loyally until the end of the expedition.

We watch the two bears all afternoon with both cameras and our binoculars, learning to interpret the thermal camera images in the remaining twilight and to identify polar bears on these systems so that we will be able to detect them in total darkness.

In the afternoon, the bears suddenly begin to move around the ship, still some distance away but moving toward the stern with increasing purpose. They get faster and faster. If they go around the ship now and appear from behind on our starboard side, our teams won't be able to escape the ice. When the bears are directly behind the ship, I decide to evacuate. I interrupt all radio traffic with the designated code— "BREAK, BREAK"—so that all channels will remain open for the urgent message to come. I look around quickly; every team has a secure

and direct route back to the ship, and I instruct them to return immediately. I wait for each team to radio and confirm that they've received
the message, then I sound the ship's horn as a general evacuation signal.

Nobody asks questions or tries to argue; across the ice, the established procedures begin. Each team assembles by their armed polar
bear guard, starts up their snowmobiles, and makes sure everyone has
space on the sleds. The teams make their way back to the ship in a quick
and orderly fashion. The polar bears continue around the back of the
ship and draw closer. After just eighteen minutes, dozens of researchers
are back on board safe and sound. The gangway is raised. An exemplary
evacuation. This is one reliable team!

And not a moment too soon; the bears move quickly from the stern
to our ice camp on the starboard side. They clearly feel far too comfortable. We need to drive them away again to prevent habituation.

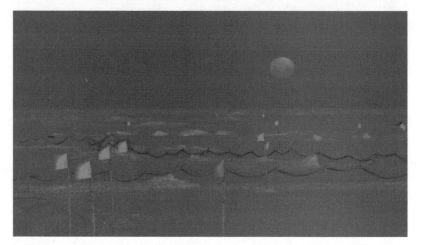

▶ The power lines and flagged tracks of the research camp in the moonlight.

I send out two expedition members on separate snowmobiles with
flare guns to drive the bears away. This is a tried-and-tested method; if
one vehicle fails, they can ride together and evade the bears. We would
proceed the same way if colleagues required assistance on the ice and
didn't have a snowmobile, or their vehicle failed to start. There are
always two people on two snowmobiles, both with flares and guns for

safety. In case of emergency, two more snowmobiles are waiting on the gangway.

The first two team members ride onto the ice, standing on their running boards to get a better view. The bears hear the engines and look over in astonishment. They seem perturbed by the noise. Bears normally take to their heels when snowmobiles approach, but we don't want to get too close. Once they're about 150 yards away from the animals, both team members fire their flare guns in the air between the bears and snowmobiles—the perfect spot. The bears immediately trot away from the ship and out of our camp, into the rugged landscape of the fortress. My two colleagues follow them a little farther and fire their flare guns again. Then they return, their mission successful. The bears continue to move away from the ship before disappearing from our field of vision. But we won't be working on the ice again today.

In the evening, I sit down with the captain. We're increasingly worried. Normally, bears roam the central Arctic, covering huge distances in search of what few seals there are. Surely our ship and its various smells aren't so fascinating that the polar bears will return again? Even today they didn't seem remotely afraid of the ship. We need to convince these bears to follow their natural lifestyle and continue across the ice, or they'll have a problem. Bears who grow accustomed to and stay close to humans are living dangerously. They won't find enough food here. And if things were to get critical, all it takes to kill a bear and save a human life is one good shot. Bears' curiosity can prove their downfall. Soon the last bit of daylight will vanish—and spotting bears in plenty of time isn't so easy in total darkness. This will become painfully apparent later on.

October 13, 2019: Day 24

THE DAY GETS OFF to a bad start. Our furry friends are sitting placidly on the ice, about half a mile to the front of the ship on the starboard side, not far from Met City. They still seem to like us, even though we've been pretty rude to them so far.

▶ Ocean City glitters in one of the *Polarstern*'s three bright searchlights.

We drive them off again with two snowmobiles and flares, then follow up with the helicopter—a textbook procedure. I hope it's enough.

In the afternoon I allow my colleagues back onto the ice and our outside work continues. As the day progresses, the temperature drops from −14°C to −25°C (6.8°F to −13°F). A fresh wind blows through the camp that makes it feel even colder. Our breath condenses as soon as it leaves our bodies, and ice crystals settle on beards and eyelashes, hats and scarves. Our faces are totally covered in ice, and we're indistinguishable in our red polar suits. It's the coldest day so far.

Despite the drop in temperature, our construction work continues apace and all huts have been erected. Today we have reinforcements; for the next while, a daily helicopter shuttle between our camp and the *Fedorov* will fly colleagues over in the morning and back at night. The large Russian Mi-8 helicopters also bring cargo that we transport to the *Polarstern* from the helipad in the fortress using Nansen sleds and snowmobiles. Our colleagues on the *Fedorov* have almost completed the station network around our ice floe; just a few smaller sites remain.

As I return from my inspection tour, the full moon rises over the ice floe. A huge, deep-orange disk, it gleams in the dark-blue, almost black sky. During the night it roams around us on the horizon. And as it moves, the ice floe changes; late at night, between 11:00 PM and midnight, we experience the strongest ice pressure so far. The ice begins to rumble all around us, and the ship shudders and shakes. I stand at the back of the ship on the working deck. Suddenly the ice breaks open next to me with a loud bang. The ice floes push into and on top of one another with loud creaks, and an ice ridge about ten feet high rises up next to me where previously there had been calm, even ice. This is probably the result of the resplendent full moon, which amplifies the tides and the internal oscillations of the sea ice. Nansen reported that the cycles of ice compression and opening were particularly intense during full and new moons.

October 14, 2019: Day 25

NO BEARS TODAY. After working several seventeen-hour days, I have canceled a press meeting that was scheduled for this evening. It's the first time in ages that I've had a couple of free hours. I listen to music in my cabin and practically fall asleep on my feet.

▶ The *Polarstern* in the initially flat MOSAiC ice floe. Later, this area will be totally reshaped by mountains of ice.

October 15, 2019: Day 26

ANOTHER WHOLE DAY without bears! It's looking good—our last attempt to drive them away must have worked.

In the afternoon I finally have time to take a good look at the camp, which is making wonderful progress. ROV City is almost ready to roll; there's an ice hole for our underwater robots, the ground around it is covered in wooden boards, and the orange tent (which covers the hole) is nice and taut. It glows beautifully in the twilight when lit from inside. The hut at the Remote Sensing Site is also finally ready. The thirty-six-foot tower is almost finished in Met City—it's currently lying on its side—and the base for the ninety-eight-foot mast has also been built. Some of the researchers helping to construct Met City have to leave on the *Fedorov* in two days' time, but it'll be completed before then. The only slight hitch is in Ocean City; the device controlling the CTD rosette winch keeps emitting an error message. The technician responsible is also due to sail on the *Fedorov* and is working tirelessly to sort it out.

Riding back on my snowmobile, I stop for a moment. The *Polarstern* glows in the twilight, around it nothing but a vast, icy expanse. The ship might be a massive hunk of steel, but from here it looks positively homey. I allow myself a few minutes to enjoy the view before continuing my work on board.

Over the last few days, more and more cracks have appeared in the ice floe—one in the corner, part of our route to Ocean City, and another moving diagonally from the front of the ship toward Met City. Plus, there's the crack in front of the bow that formed soon after we broke into the ice floe and has continued to open and close ever since.

But that's nothing compared with what happens tonight.

At around 4:00 AM, my cabin phone rings and I get an alert from the bridge. Even in my room, I can hear and feel the ice rumbling and crunching around us, making the ship vibrate. I quickly pull on my polar gear and rush to the bridge, where the guards explain what's happening; the ice has started moving and is pushing the *Polarstern* forward. In front of the bow, our crack is rising into a three-foot-high pressure ridge and the mass of ice is rolling toward the cable that leads to ROV City. I can't do anything about that right now, but things are precarious on the starboard side, in our loading zone right next to the ship; there's a lot of equipment there and we need to act fast.

I hurry to the working deck, where the cargo officer and a ship's mechanic have already lowered the gangway. Outside, the noises from the ice are indescribable, not of this world. Loud banging, cracking, and crunching, screeching and moaning. I've never experienced such extreme ice compression. Below, the ice has pressed the side of the ship against the floe and then loosened its grip, leaving the front of our loading zone in ruins. One box and some timber are already floating in the water, while another box of equipment threatens to topple into a crack. Another fissure is quickly opening crosswise under the row of snowmobiles parked on the ice overnight. One of the vehicles is already hanging half in the water, its runners wedged under the floe, while others are in danger of disappearing into the crack. Our expedition would be lost without snowmobiles!

I quickly check what's happening, then sprint down to the snowmobiles with the cargo officer and a member of our logistics team, who has been woken on my request. We start the still-mobile vehicles and maneuver them out of the danger zone. Then we join forces to retrieve the half-sunken snowmobile; we manage to get it afloat and drive it away. In the meantime, the cargo officer has pulled on a survival suit and started to retrieve boxes and wood from the water. Just in the nick of time!

After dealing with the immediate risks, we pull any other equipment on the ice toward the cable path in front of the ship. We were wise to run the cable along a solid ridge of last year's ice, which is thicker and more stable.

The salvage operation takes hours. We return to the ship, satisfied with our efforts. I discuss the situation with the captain and team leaders immediately. Night turns to morning without pause.

▶ HOW DO MOUNTAINS FORM IN SEA ICE?

The sea ice that covers the Arctic Ocean isn't rigid; it is subject to many forces. Wind chafes its rugged surface, transfers momentum, and drives the ice across the water; tides and currents set it in motion; the tidal range raises and lowers the surface of the ice; and after a big storm, the waves on the open sea can be measured more than sixty miles behind the ice edge. These processes can cause enormous pressure to build up in the sea ice, and this pressure needs an outlet. The ice breaks open and the ice floe fragments warp, push onto and beneath one another in huge chunks, or rise up vertically. This creates "mountains" over ten feet high with ridges and jagged peaks. These pressure ridges can extend along the ice for miles. The biggest ridge we saw during the MOSAiC expedition was over eighty feet thick.

▶ The ROV City power line is swallowed by a pressure ridge and has to be freed.

October 17, 2019: Day 28

YESTERDAY, ONCE THE ICE calmed down, we cleared up the damage. The last cargo boxes were helicoptered over from the *Fedorov* and lifted aboard. Now it's time to say goodbye to some of our companions and welcome many new faces from the *Fedorov* before she heads for home.

From 8:30 AM, the leavers assemble in groups on the working deck; the Mi-8 can only carry about twenty passengers with kit bags and luggage. Every two hours we use the Nansen sleds to transport a group to the helipad inside our fortress, where the Russian helicopter collects them and flies to the *Fedorov*.

In the afternoon we gather on the helipad in the fortress for one last meeting with our friends. I am joined by the two captains, Stefan Schwarze and Sergej Sidorov; Vladimir Sokolov of the AARI; Thomas Krumpen, the cruise leader on the *Fedorov*; and a few others. We ride snowmobiles from the *Polarstern* to our arranged meeting point, where we are met by the Mi-8. It's a bizarre situation: the *Polarstern* on one side of the horizon, a little bright spot in the night, and the *Fedorov* on the other. We wait between the two, in the middle of nowhere, waiting for a Russian helicopter to descend in the central Arctic polar night.

When it arrives, we greet each other one last time. After successfully putting everything in place for our research, we're feeling festive. We drink whiskey from thick glasses with ice broken straight off the floe. There are toasts and plenty of laughs. We couldn't have done this without our Russian friends, and we make sure they know that. We are happy as we bid each other farewell. The Russian delegation boards the helicopter, nearly knocking us over with the downwash, the strong draft created as the huge helicopter flies away—another good laugh for those of us left on the ice. We return through the night to the *Polarstern* in the best of spirits.

October 18, 2019: Day 29

THE *FEDOROV* LEAVES today. Visibility is good, the Arctic allowing us to see our friends one last time. It'll be months before we glimpse another human in this rugged, alien world. For a good two hours, we watch the *Fedorov* getting smaller and smaller in the dark-blue twilight, its lights getting weaker and weaker. Then she's swallowed by the horizon. We are alone.

Part II

WINTER

▶ The *Akademik Fedorov* has gone, and the team are alone in the vast Arctic and dark polar night. The next human settlement is nearly a thousand miles away.

4

ALONE AT THE END OF THE WORLD

October 24, 2019: Day 35

FOGGY TO START, then it brightens up. It's still warm, around −14°C (7°F).

The *Fedorov* left almost a week ago. We've been alone ever since, thousands of miles from home. And our icy surroundings present new challenges every day.

The thick bow lines fixing the front of the *Polarstern* to the floe are under immense strain. They could snap at any time or rip out the heavy ice anchors, welded steel joists frozen in drill holes that can withstand tremendous force. Now, however, the ice pressure at the front of the ship threatens to exceed anything a human-made anchor can withstand. We are no match for the forces of nature.

Strained to breaking point, the lines present a deadly threat: if an anchor bursts out of the ice, the line will catapult it toward the ship like a slingshot. We quickly add another anchor behind each of the steel anchors and use them to secure the main anchor with loose safety lines, which will stop any projectiles in midair should the worst happen. But even if the anchors hold, a line might tear and destroy everything in its path as it ricochets. We decide to close off the whole area. Not a day goes by without me reminding the team that these lines are extremely dangerous—nobody is allowed to enter the area, ever.

The stern lines aren't all that great either; they're only anchored in one little floe fragment and won't hold for long. Our stern is no longer fixed to the ice floe. We desperately need the temperature to drop. The cold is our friend; to increase stability, we need the ice to grow.

Just after the *Fedorov* left, the ice started to move again, rumbling and crunching. ROV City (and our underwater robot) shifted a whole six hundred yards to port in a huge shear zone off our bow. The ROV is one of our most valuable research devices; we can't afford to lose it! Luckily, a team managed to recover the robot, and its computer hut, using a helicopter. We've already chosen a new location for ROV City, nearer the ship and straight ahead; a team will head out there today to dismantle the tent, cable, and whatever's left of the old station and bring them back to the ship, ready for rebuilding.

We've also had a few polar bear visits, including a large male who gave our camp a thorough inspection. He decided to take a closer look at Balloon Town, where the 770-pound hangar tent was unpacked and awaiting construction—he threw it around as though it were nothing more than a backpack. We patch the lacerations he left in the tent, which will remind us of his visit for the rest of the expedition. Another bear used one of the long flagpoles we placed on the ice for orientation to pull himself onto his hind legs. He clearly wanted to get at the orange flag fluttering enticingly in the wind, thirteen feet off the ground. Our photographer, Esther Horvath, managed to capture the scene, and will go on to win a World Press Photo award, the most prestigious accolade for press photographers.

▶ One of the two *Polarstern* helicopters rescues ROV City, displaced by ice shearing on October 19.

And sometimes things just break, like the FIRST navy, the sophisticated 360-degree thermal imaging camera in our crow's nest. It's crucial for spotting bears in the polar night. We installed our replacement camera straightaway; sadly, that too will prove no match for the harsh Arctic conditions. Losing this camera will make our lives difficult now that we have entered total darkness.

No matter how well we plan and prepare, there's no such thing as normal in this environment. Unforeseen events just keep on coming. And we just keep on adapting. Despite all the experience we have gained on previous expeditions, the ice is difficult to anticipate. But this is what makes it so fascinating. And even though we're ensconced in an icy wilderness, a freezing wasteland, there's no chance of getting bored. Not only because there's always something to do on board or in the research camp, but also because the Arctic itself is constantly evolving, presenting us with new challenges. The ground beneath our feet may look solid, but it isn't—there's over thirteen thousand feet of ocean down there. The wind, snow, and movement of the water with the tidal range and current are constantly altering the face of the ice. From one day to the next, a big chunk of ice covered in icicles and snow crystals might be replaced by a snowdrift or a channel of open water.

We've established some routines on board. Anyone with time to spare meets for a game of table tennis in the lower cargo bay. To make it a little more exciting, we hold a daily tournament; the last player standing gets to wear a medal that Lutz Peine, one of our officers, has fashioned out of cardboard. Anyone who scores a hat trick—winning three days in a row—has to play with their nondominant hand. Lutz is the first to fall foul of this rule. We often play a game after lunch, and today I've made it to second place—a rare success!

Every Sunday, we hold a weighing club—it's a tradition on the *Polarstern*, and always great fun. In the morning we all meet in the machine shop, where an ancient beam balance hangs from the ceiling with heavy counterweights. We take turns standing on a wooden board and hanging on the scales, to the amusement of the onlookers. We then try to guess whether we'll be lighter or heavier the following Sunday. Our guesses are usually close, but not quite right; strangely enough, the scales frequently contradict our predictions, even if we're certain we've gained or lost weight in the last week. There's a rumor that the master of the scales might be meddling with them. But it's fine; if you guess wrong, you put a little cash in a jar and all the money collected at the end of the expedition goes to a children's hospital. So who's going to argue?

These rituals and leisure activities are just as important as our daily schedule of meals and work. We've left civilization behind and have no connection to the rest of the world; since the endless polar night swallowed every natural rhythm, we can't even use the normal cycle of the day for orientation. And it's important that we remember to have fun. Even in Nansen's day, explorers knew the importance of combining regular work with social activities. It helps us to stay healthy—both physically and mentally—over such a long period and under such extreme conditions.

▶ Met City at the edge of the fortress on our MOSAiC ice floe.

October 27, 2019: Day 38

A QUIET MORNING. Weighing club takes place at 10:30 AM, and for once I have gained weight, over a pound since last Sunday. The carpenter replaces the water valve in my room. Yay, finally I can adjust the temperature of the shower! I wash my clothes. I haven't had enough sleep.

The sky is beautifully clear today. At midday, with no clouds, we can see a very weak reflection of the sun to the south, deep below the horizon. The temperature has dropped in the last few days. At −26°C (−15°F) with no wind, the lowest atmospheric layer becomes a flurry of ice crystals. We can see it and feel it, the whole atmosphere radiates it: winter is near.

On November 14, 1894, at 82° north, 114° east, Nansen writes:

The ice-fields stretch all around, bathed in the silver moonlight; here and there dark cold shadows project from the hummocks, whose sides faintly reflect the twilight. Far, far out a dark line marks the horizon, formed by the packed-up ice, over it a shimmer of silvery vapor, and above all the boundless deep-blue, starry sky, where the full moon sails through the ether. But in the south is a faint glimmer of day low down of a dark, glowing red hue, and higher up a clear yellow and pale-green arch, that loses itself in the blue above. The whole melts into a pure harmony, one and indescribable.[2]

What more is there to add? It's astonishing how much Nansen's description resonates today, and how vividly he conveys the ambience! Nansen saw exactly what we are seeing now, the same sky, the same ice all the way to the seemingly endless horizon, the same fascinating remnant of Arctic twilight before night descends over the polar cap. Where else has the landscape escaped total alteration by humanity, where else are there no lights, no aircraft in the sky, and no human structures, just as Nansen experienced over 125 years ago?

But appearances can be deceptive; this frozen landscape will not stay the same forever. The ice beneath us is only half as thick as it was in Nansen's day. Our generation may be the last to experience an Arctic that is covered in ice all year round.

October 29, 2019: Day 40

WE MAY BE surrounded by nothing but night and ice, but there are higher lifeforms beneath us. A few days ago, we realized that beneath the ice, the water is teeming with fish! A sensitive sonar picks out the fish and shows us exactly what's going on down there. While day and night were still distinct, the fish stayed at a depth of almost a thousand feet during the day and ascended to five hundred feet at night. They do this to prevent predators from finding and eating them. Now that it's dark all the time, they stay at five hundred feet. This suggests that vertical migration in the ocean is controlled by a light sensor—and not by a light-synchronized body clock! Otherwise, the fish would continue to ascend and descend, and their movements would simply become less regular.

Yesterday we dropped a longline into the water. If we're lucky, we'll be able to identify a few specimens more precisely, rather than just watching them on the sonar.

It's a big day for the anglers. We haul in yesterday's long fishing lines. Nicole Hildebrandt from the AWI pulls up the line hand over hand. She says she can feel the fish already. Everyone laughs; nobody believes her. But then a wide-eyed face appears in the ice hole. And another! Our anglers can't be convinced to take these magnificent fish to the kitchen. They are for research purposes only.

In fact, all our research is gathering speed. Instruments have been set up, test runs have been completed, and standard operations are now beginning. During our years of planning, we drew up weekly schedules that set out precisely where each team would be and when—for example, when measurements would be taken on each snowfield, how often we would fly to the larger floes in our station network, and when the CTD probe and other large ocean instruments would be lowered into the water. Days are long here; we want to obtain as much data as possible on this unique expedition. Yesterday, one team held its first coring day, and very successfully too, returning with fifty ice cores drilled from the floe. Backbreaking work for a single day and an amazing achievement in these conditions; it's getting colder and trickier out on the ice!

The Ice Cycle—and How It's Changing

FALL AND EARLY WINTER is the time of year when the ice freezes. Freezing and thawing processes have a huge impact on the ice and the climate system as a whole. But we still don't fully understand them. In particular, we lack information on how these processes affect the radiation under, in, and on the ice, and how they affect the Arctic's energy budget (the balance between all processes supplying energy to and removing it from the Arctic system). But these are the processes that determine how the ice in the north interacts with the ocean and atmosphere, and how this impacts the ecosystem. Thanks to climate change, these processes have undergone some extremely rapid changes.

The Arctic Ocean has an annual cycle. As soon as the sun sinks deeper, new ice begins to form and existing ice consolidates. We arrived during this phase; new sea ice is forming, and the melt ponds that developed in the summer are already covered with ice and barely visible to the naked eye. But the ice doesn't really start to thicken until the sun has totally disappeared, and we have now reached this point. The upper ocean layer gets so cold that the ice grows from below. Depending on its salt content, sea ice starts to freeze at around −1.5°C to −1.9°C (about 29°F). As the ice grows thicker and is covered in more snow, this freezing process slows down. Ice and snow provide the ocean water with more insulation from the freezing-cold atmosphere, even as the temperature drops below −40°C (−40°F). Once the ice is around six to seven feet thick, this "thermodynamic freezing" slows down and almost grinds to a halt. For the ice to become thicker still, ice masses must push against one another and form thick ridges; the cold and pressure then force these ridges to freeze together and form a solid unit. This is called dynamic secondary ice growth.

One-year-old ice looks more even, and its flat sections are rarely much more than seven feet thick; it has had only a short time to push on top of other sections. Multiyear ice is totally different; it has already undergone several cycles, and its surface is covered in high, rounded humps—aged and eroded pressure ridges—meaning that it can be far over ten feet thick. Our floe's solid core consists of two-year-old ice. In the past, vast areas of the Arctic were covered in thick multiyear sea ice. Today this is the exception, rather than the rule. Like almost all the ice around us, our floe consists mainly of thinner, relatively even one-year ice or new ice that is just starting to form. We have an invaluable opportunity to study climate processes on all these types of ice.

When summer comes and the sun rises high above the horizon, shining down on the ice twenty-four hours a day, a new process begins—thawing. This transition takes place between late May and mid-June, depending on the region. The ice's bright surface gives it a high albedo, or ability to reflect radiation—covered in fresh snow, it can sometimes reflect more than 80 percent of the sun's energy back into space. What remains is sufficient to warm the ice and melt it partially or completely. Melt ponds develop and the ice edge retreats. In the central Arctic, some of the new ice that formed the previous year might survive the summer.

Next winter, the surviving ice floes and ridges can repeat the process—growing, forming hills, and so on—until strong multiyear ice develops over time. This is different from the sea ice in most of the Antarctic, which tends to melt completely every summer and form anew in the winter.

But, as we have seen, thick multiyear ice is increasingly rare in the Arctic. Instead, it is now dominated by thin one-year ice, which often simply melts again the next summer. This is due to global warming.

November 2, 2019: Day 44

THERE HAS BEEN no trace of daylight for some days now. It's −22°C (−7.6°F) and calm. We set up the large hangar over at Balloon Town, home to Miss Piggy, the red tethered balloon that takes atmospheric measurements. The orange tent is about twenty-five by thirty feet and almost thirteen feet high. It's the camp's largest piece of infrastructure. But first it needs to stand.

▶ Balloon Town with Miss Piggy, our beloved tethered balloon, in its parked position. The balloon can float up to 1.2 miles into the atmosphere and take measurements above the ice.

We've chosen a calm day, which makes our work easier. The enormous hangar is held up by a framework of tubes that will be inflated by two compressors. The hangar has no solid structure, making it highly robust in even the worst of storms—there's nothing to break. Over the last few days, we have hacked into the ice to create an even surface and screwed together wooden planks to make a stable floor. Now the tent package—the same one that recently proved so interesting to a polar bear—stands in the middle of the floor. It's still hard to imagine this as a spacious hall. We roll it out quickly and attach it to the compressors. There's no time to lose—in the Arctic, the weather could change completely from one hour to the next.

Even in such low temperatures, we soon find ourselves sweating, toiling in a cloud of our own breath with no wind to blow it away. As I attach the bottom of the hangar to our floor panel, I briefly hold a screw between my lips. Not a good idea—it sticks fast and I struggle to pull it off.

We turn on the compressors, and nothing much happens; our thin power line can't cope. We calculated the power required for Balloon Town based on the continuous output of the compressors, but our plans are foiled by the amount of electricity required to start the motors. We quickly procure some gasoline generators, but even they

▶ Even at –30°C (–22°F), working on the ice soon makes you sweat. The vapor escapes through the collars of our polar suits. When the air is still, we're soon surrounded by clouds of our own breath, which refract the light from our headlamps and dazzle us.

85

can't get the compressors running. So we run a thicker power line from the Ocean City distributor around a hundred yards away. Now we have enough power; the compressors start and in no time at all the huge hall unfolds, as if by magic.

I end the day by walking around Balloon Town by the light of my headlamp. The ice sculptures in the fortress seem to be the creation of an unknown artist. They would certainly be welcome in any museum, but instead they stand alone in the vast, pitch-black Arctic, ordinarily hidden from human eyes. Everything is covered in a thick shield of hoarfrost made from large, glittering ice crystals. The ship glows in the distance, tiny and safe and warm, our home in this utterly fascinating, yet hostile environment.

▶ The world's most northerly ice bar, site of extraordinary moments amid sweeping sea ice, all alone near the North Pole through months of polar night.

Back at the ship, a team is drilling an ice hole for the large CTD rosette and the biologists' nets. There's already a small CTD rosette in Ocean City, but it doesn't provide the water depths and quantities we require. We need an ice hole that measures at least thirteen by thirteen feet and is right by the ship, within reach of the onboard crane to which we will attach our instruments. The team has worked strenuously to cut the ice into smaller blocks, which still weigh several tons. Chainsaw Alli is in her element, brandishing the long-bladed saw, while others use ice core drills to create a series of holes around the contours of the ice blocks, gradually separating them from the rest of the ice. We've also made a hole through the center of the middle ice block; we push a wooden beam on a line through the hole and twist it to create an anchor. Now we need the crane to remove the enormous ice block—it probably weighs around two tons. But can we do it? Nobody has ever constructed a hole like this under these conditions. The crane increases the tension, and then there's a jolt—to the cheers of the spectators, the ice block hangs in the air. What a relief, it really does work! We use the same method to remove the remaining ice blocks from the hole until this elementary piece of research infrastructure is complete. Our large oceanic instruments now have access to the water column beneath us!

We clear the fresh new hole of the slush ice created by the drilling, then we lower an oversized electric whisk into the water so that it doesn't simply freeze again. The crane deposits two of the ice blocks on the floe near the ship. We thread colorful lights through the drill holes and voilà, a glowing ice bar! It will serve us well for the rest of the winter.

After the long days of toil, it's important for the team to celebrate together. We held a Halloween party a few days ago, a feast of creativity and improvisation. One colleague transformed himself into a perfect replica of C-3PO, the clumsy robot from *Star Wars*, skillfully whipping up a costume from survival blankets and anything else he could find on board; he was accompanied by R2-D2. He even had the movements down to a tee. Another team member appeared to have suffered a gruesome accident, impaled through the head with one of our flags, blood and brain matter dripping down his body. It was so realistic that we had to resist the urge to carry him to the *Polarstern* hospital.

▶ TIME AT THE END OF THE WORLD

The planet is divided into twenty-four time zones based on lines of longitude. They meet at the North Pole where, theoretically, you could walk across every time zone in just a few steps. Time loses all meaning here. When you get this far north, the rhythm of day and night is suspended; the sun rises and sets just once a year. As we traveled on the *Polarstern*, we reset her clocks twelve times to match her geographical position. But once we were far north we could essentially choose our own time. During the weeks after we docked at the ice floe we gradually shifted the clocks to run on Coordinated Universal Time (UTC), which is based on the Greenwich meridian and is the standard in London. This makes it easier to communicate with the rest of the world and to convert the time stamps on our data.

We also have a reason to celebrate today: we're exactly halfway through the first expedition phase! That is, as long as we don't count the long journey back (nobody can say how long that will take). We hold a barbecue outside on the working deck and celebrate in the adjacent working rooms, the wet lab, and the friction room. The friction room is where the winch cables run from the *Polarstern*'s belly to the outside of the ship; we also make sure it has tables and benches so that we can sit together at the end of the day.

Parties on the *Polarstern* are very distinctive. How often have I danced into the night on previous expeditions, like the ones in summer in the Antarctic? The darkened party room almost makes you forget where you are; when you step outside for a breath of fresh air, you're overwhelmed by the sight of icebergs drifting past, glowing pink in the blazing midnight sun, or the ship's shadow on the flank of a broad, flat iceberg.

This party will be held in constant darkness, with no distinction between day and night. By this point, it's clear who's getting along well—and who's getting along extremely well. An experience this intense with such captivating sights forges very close relationships. We've all seen it happen—and it doesn't get much more intense than a polar expedition. I've made friends for life and know quite a few couples who've gotten together on expeditions, started families, and are raising the next generation of polar explorers.

The party continues, developing its own special dynamic—like all good parties. We even gain an hour when the clock is reset. We dance until 3:00 AM. Twice.

November 3, 2019: Day 45

A QUIET MORNING after the party. It's windier today, twenty-two miles per hour, from the southeast again.

At around 2:00 PM, one of the trip wire flares goes off. We set up the wire to warn us of polar bears approaching in the darkness. We check the wire but don't find any animal tracks. The flare was probably triggered by the wind and the ice on the wire.

November 4, 2019: Day 46

ONE OF OUR COLLEAGUES has frostbite. Instead of donning his thick mittens and using the stylus to operate the navigation equipment, he used his fingers, clad in nothing but thin gloves. Now the onboard doctor has to treat his frozen finger. It's some time before we know whether it can be saved. Thankfully, it can.

This isn't good. We can't allow errors like this to creep in. In our daily meetings, I remind everyone to watch out for their colleagues at all times, to warn anyone who is inadvertently exposing their skin to the wind, and to constantly ask others how they are and whether everything's okay. In this environment, it's crucial that we look after ourselves and one another. In case of doubt, we must suspend or abort our work, even if it eats into our time. I want us all to return from the Arctic with the same number of fingers we had at the start.

▶ A FLOATING HOSPITAL

Despite all precautions, emergencies and serious illnesses are always a possibility on expeditions like these, which is why the *Polarstern* has a small (but fully functioning) hospital with an operating theater, a treatment room, a sick room, an isolation ward for infectious conditions, and a large supply of medication. The onboard doctor and nurses can stabilize open fractures using a fixator and can even perform simple operations, for example on an appendix or hernia. The doctor also has an ultrasound machine and a digital X-ray and can run blood work. In the second half of the expedition, which coincides with the COVID-19 pandemic, coronavirus diagnosis equipment is brought on board by one of the supply icebreakers. Thorough treatment options are particularly important on MOSAiC; in some phases of the expedition, it would take weeks to evacuate someone to a hospital on land. Thankfully, most of the conditions treated on board are fairly minor, such as cuts. One team member comes down with the flu and is taken to the isolation ward, which has its own ventilation system; without this, viruses could quickly spread throughout the ship.

November 9, 2019: Day 51

POLAR BEAR ALERT! Today we find ourselves in the critical situation we have trained for so often, and that we all run through in our minds. We can only hope that, when it comes to the crunch, we make the right decisions.

A bear sneaks into the middle of the camp without anyone noticing. He comes around the southwest end of the trip wire and emerges from the darkness in Met City. By the time the guard spots him, he is only around fifty yards away and exhibiting dangerous behavior—raising his nose and swinging his head from side to side, sucking in air and moving closer. This is the behavior of a bear curious to find out what's in front of him and whether a quick attack might reward him with a meal. This isn't a training exercise, this is deadly reality, and the bear straight ahead of us isn't made of cardboard; he's a predator who weighs half a ton and can run like the wind. If the bear attacked from this distance, he would be on us in about three seconds. Polar bears can cover short distances in enormous bounds and reach up to twenty-five miles per hour. The polar bear guard, one of our most experienced, reacts immediately. This close, there are only two (equally poor) options. You can fire a flare into the snow between you and the bear, in which case the cartridge will detonate under the snow with a muffled *pfff*. Alternatively, you can fire very high, almost directly above your own head, in which case the cartridge will detonate high in the air. The one thing you absolutely must not do is fire in the bear's direction! If you do, the cartridge will fly past the bear, detonate behind him, scare him, and send him running toward you. It's a classic lose-lose situation; when the bear's this close, there's really no right answer.

The guard decides to fire high. The cartridge explodes way above the bear's head, and he has no idea where the banging is coming from. He looks up but doesn't fall back, immediately returning his attention to the people on the ice; he probably hasn't realized that they are the source of the noise. The guard reloads and fires high again. The effect is the same; the bear doesn't retreat.

This is the greatest of emergencies—the bear is even closer now, hasn't been frightened by the flares, and continues to approach. Lives are at stake. The guard follows protocol. He loads his gun and aims. In a last attempt to save the bear's life, he fires above his head, trying again to drive him away.

The bear realizes that the humans are the source of the noise. Things are getting too scary for his liking, and he gallops off into the darkness. He leaps nimbly over the trip wire—which is supposed to protect us from visitors like him—without setting it off. From the bridge, the ship's searchlights follow his progress.

The bear doesn't run far, but it buys us enough time to evacuate the area and get everyone off the ice. When the message comes through that everyone's on board, I sink into a chair in relief.

Now we need to make sure that the bear keeps moving and doesn't take up residence. A second, armed team member is sent onto the ice to support the polar bear guard; they follow the bear on snowmobiles and fire a few more flares from a safe distance. From here, the flares have the desired effect, detonating in the air between the snowmobiles and the bear, who continues to flee. This moves him sideways out of our scent stream so that the smell won't entice him back to the ship. But even this doesn't frighten him all that much; he keeps stopping and turning back toward my colleagues. He raises his nose and sniffs, still interested in the potential source of food. Each time, the flares let him know that it wouldn't be safe to attack. Eventually he moves on, and the two snowmobiles return to the ship. We raise the gangway. The situation is stable for the moment.

We spend another half hour on the bridge, watching the bear with our binoculars, searchlights, and the older thermal imaging camera, which is still working.

▶ Hans Honold climbs a pressure ridge to watch for polar bears. Every team on the ice has a polar bear guard; expedition members serve as bear guards in shifts of two to three hours.

The bear glows on the thermal imaging screen. He's agitated and tired and needs to release some heat. He rolls around and sticks his head in the snow; his head is less insulated than his furry body. It's remarkable to watch a polar bear cooling off in the snow in strong winds and temperatures of −25°C (−13°F).

The bear keeps moving until he disappears into an area obscured by pressure ridges. Via thermal imaging, we watch him hide behind a ridge and poke out his warm head to look at the strange thing that's just given him such a fright. The feeling is mutual. Eventually we can't see him at all, and his trail disappears into the darkness.

Since I don't know where the bear is, and he was clearly interested in our camp, I keep everyone off the ice for the moment. The gangway stays up. The bear didn't look well nourished, and his behavior suggests he was desperately searching for food. I'm worried about him, and rightly so—less than an hour later, we spot him abeam on our portside. He shows up on the thermal imaging camera, which doesn't provide a 360-degree view (like the broken FIRST navy), but constantly monitors port astern and the area behind the ship. Over the next thirty minutes, he walks around the ship, maintaining a distance of a few hundred yards, continually seeking cover behind pressure ridges—classic hunting behavior. But he's clearly uneasy; he's probably keeping his distance because of our efforts to drive him away.

The bear rounds the ship's stern, almost reaching the end of the trip wire on our starboard side. Before the end of the wire, he finds a GPS station, marked with a flag and positioned here to provide coordinates for our work on the ice. This buoy has already endured a few polar bear visits and doesn't escape this time either. The bear examines it thoroughly and pulls himself up by the flag.

He's far too comfortable for our liking and growing more familiar with our structures. Just as I'm debating whether to send out the snowmobiles again, the bear abandons the GPS station and walks to the trip wire. It's obvious that he can see the thin wire perfectly well in the darkness. He sniffs it cautiously and follows it for a few yards. What's going to happen now?

PING!! The trip wire charge goes off. A rocket flare soars into the sky just fifteen feet from the bear, lighting up the whole area as it floats on a parachute. The bear leaps away from the wire and dashes off. This time, we don't need to intervene. The bear trots away and is never seen again. But for the next few days, we are haunted by the prospect of this humongous animal lurking in the dark.

November 10, 2019: Day 52

IT'S SUNDAY. We can tell because there's eggs for breakfast. Something went wrong when we loaded the food onto the ship, so we've had to ration breakfast eggs to Thursdays and Sundays. We're also running low on vegetables. Lunch is often served without cauliflower or broccoli. We all look forward to vegetable days and hope for fresh supplies when our supply ship, the *Kapitan Dranitsyn*, comes in mid-December.

The ship is much quieter than usual this morning. There are no plans to work on the ice and many expedition members are taking time for themselves, enjoying the rare opportunity to relax for a few hours.

I'm in my cozy cabin, playing my favorite Sunday-morning music, the same I enjoy at home on the weekends—*Soul Lounge* by Bona Fide. As I listen, I think about yesterday's incident.

It was seriously bad. If we don't spot a bear until it's just fifty yards away, then there's always a danger of things going wrong. Just a few seconds more and the bear might have been in the middle of Met City, among all the people, which could have been a catastrophe. I recall the last polar bear attack on Spitsbergen. A bear broke into a camp unnoticed, killing one person and seriously injuring several others before they were able to shoot it. I shudder just thinking about it.

In the darkness, with nothing but headlamps and searchlights, there's really no way to ensure that we'd quickly spot a bear in our fortress, where the views are obscured by ice blocks and high ridges with deep recesses in between. Our FIRST navy thermal imaging camera would have allowed us to spot an approaching bear much sooner from the bridge—it helped with every previous bear visit—but now the camera has stopped working.

▶ **WHAT DO POLAR EXPLORERS EAT?**

The *Polarstern* carries 1,500 different kinds of food and supplies for the first three months of the expedition alone: 14,000 eggs, nearly 400 gallons of milk, a ton of potatoes, 150 jars of chocolate spread. Then there are two containers only to be accessed in an emergency, with another two months' worth of long-lasting foodstuffs. It's important to eat well on polar voyages. Meals give structure to the day, and good food helps maintain a positive atmosphere on board. Two chefs and two kitchen assistants work from 5:00 AM to 6:30 PM to feed the crew of around a hundred people. They can peel a hundred pounds of potatoes for just one meal. The baker starts work at 2:00 AM so that everyone can enjoy fresh rolls daily and even an afternoon cake. The *Polarstern* usually serves traditional German fare such as goulash, currywurst, and schnitzel. As on every voyage, we eventually start to run out of fresh vegetables. The lettuce is the first to go, followed by tomatoes and cucumbers. Some team members find themselves dreaming of broccoli and cherries. By the end of the first phase in December, the crew will have eaten 12.7 tons of food, more than on most other expeditions. We need to consume a lot more calories than usual to work on the ice in such low temperatures.

Essentially, whoever's on polar bear duty has nobody to rely on but themselves. It's an immense challenge: total concentration for two to two-and-a-half hours, staring into the darkness with just the beam from their headlamp, seeing nothing for weeks and yet maintaining 100 percent focus every single second to act in a fraction of a second if a bear approaches. Everyone has to take their turn. But this approach no longer seems safe enough to me. We need to extend the trip wire around the whole research camp; at least then, a polar bear will be much less likely to surprise us at work. We need to get this sorted right away.

In the afternoon, I ski along the trip wire to take a closer look. From the tracks, it seems that the bear simply leaped over the wire as he fled without setting it off. Snowdrifts had closed the gap between the wire and the ground. The wire needs to be checked regularly and raised if the snow builds up.

Then I carry on skiing north. It's almost −30°C (−22°F) with practically no wind; I can just feel a slight drift from behind. The only thing I can hear is my skis gliding across the cold snow. They sound much different than on warmer snow—harder, scratchier—and the sound carries farther on the crusted surface. The sound alone is a good indication of the temperature.

The orange-tinged full moon is low on the horizon, huge and unreal, the stars glittering above it in the black depths. I glide through the darkness between huge pressure ridges, past epic ice sculptures, and over low-lying flat plains, last summer's melt ponds that have now frozen solid. If I keep still, all I can hear is my own breathing. Otherwise, the silence is all-consuming. The moon bathes everything in a sallow light. Even without my headlamp, I can see the contours of the icy landscape. It's out of this world, with deep-black shadows in the lower-lying areas, an almost unreal, absolute black. Now and then I hear our sodar chirping quietly in the distance; it emits a loud tonal sequence into the atmosphere to measure wind and turbulence. I feel as though I've discovered a celestial body, frozen in perpetual darkness, like one of Saturn's moons or a bizarre planet from a fantastic tale. A celestial body with a neighboring planet, a low-lying orange disk that slowly wends its way along the horizon. It seems pretty far removed from the Earth we know.

The slow wind drift from behind means that my breath cloud keeps pace as I ski, and soon I am traveling in my very own cloud of freezing fog. It refracts my headlamp and dazzles me. Again and again I stop, wait for the fog to dissipate, climb an elevated point, and check for polar bears with my headlamp. This might feel more like a dead, extra-terrestrial star if there weren't hungry creatures roaming the darkness.

November 11, 2019: Day 53

WINDS AT LEVEL 5, −26°C (−15°F), snowfall, and dense snowdrift. It's a chilly day for our coring team, who today will be collecting dozens of ice cores, analyzing some of them in the field, and storing others in the ship's cold-storage rooms for their labs back home. After working on the ice for seven hours, they return to the ship totally unrecognizable, their eyelashes heavy with ice, the rest of their faces concealed behind frozen masks.

November 13, 2019: Day 55

THE TRIP WIRE FLARES have been triggered twice in the last few days, but we haven't found any polar bear tracks. In the morning, Thomas Ster-benz (who is part of the logistics team) and I explore potential routes for the PistenBully. Thomas has just spent a winter at the Neumayer Station in the Antarctic; when we met there, I realized that his many skills would be an asset for our expedition. Now he has swapped one end of the world for the other, working as a mechanic on our vehi-cle fleet and driving the PistenBully. Thomas, his colleague Hinnerk Heuck (who has also worked at the Neumayer Station and will be in charge of the ship's engines in later phases), and their fellow mechanics muster their vast experience and unbelievable talent for improvisation to keep the expedition going, even in the most difficult of circum-stances. We wouldn't be able to do anything without them. If we need a replacement part that isn't on board, they find a way to improvise. If the fuel freezes in the pipes due to its low quality, they find a solution. By the end of the expedition, the PistenBullys and many other vehicles are awash with makeshift solutions and ingenious stopgaps. But they're still working!

▶ In the impenetrable black of the night, the deep shadows cast by the headlamps, and the pale, black-and-white contrasts, the camp often gives the impression of a research station on a deep-frozen alien planet trapped in perpetual darkness.

With its milling machine and crane, the PistenBully—a huge tracked vehicle—weighs sixteen to seventeen tons and can only travel on ice over three feet thick; any less and it might not bear the load. This is why it hasn't left the ship yet, but we need it for difficult tasks on the ice floe. In particular, we need it to build a landing strip, which will be essential in spring when we plan to exchange goods and people via aircraft. At that point, the ice will still be too thick for an icebreaker to make it so far north.

There's a potential route for the PistenBully alongside the ship, where the ice is consistently around 6.5 feet thick. However, the former melt ponds still only have about twenty-eight to thirty inches of ice (they started with about half that thickness), so it can't drive over them yet. We decide to leave the PistenBully on deck and wait a while. But we do try out its milling machine, which we developed specially to mill pressure ridges on the ice plain. It's exceptional. Just as we hoped, it devours every hump of ice, spitting out huge piles of ice dust—and it's pretty fun, too.

▶ The *Polarstern* in the polar night.

The anglers have also been successful: armed with just a fishing rod, one colleague pulls a thirty-inch cod out of the Moon Pool from a depth of 1,300 feet. The Moon Pool is a well in the middle of the *Polarstern*, 36 feet deep. Anglers like to use it while the ocean is covered in ice, when a lot of effort is required to stop fishing holes from immediately freezing.

In the evening we all play soccer on the ice, the crew and scientists together. Our "Laptev Sea Stadium" doesn't require much setup: flags for goalposts, our big LED construction spotlights for floodlights, and, of course, two polar bear guards. The world's most northerly soccer match needs its armed guards. The match is as passionate as any other, as though we've forgotten where we are and the conditions in which we are playing.

November 15, 2019: Day 57

I SKI PAST ROV CITY, first along the road and then farther out into the darkness. The full moon is behind me on the horizon, bathing the white landscape in a surreal, pale light. The towering pressure ridges are bright strips between the plains, which are cast in deep-black shadows; bizarre ice formations are petrified in the wan light. The light of the *Polarstern* grows smaller and smaller behind me.

There's no wind and it's relatively warm at −15°C (5°F). My skis glide easily across the warm snow, their sounds muffled compared with those in the usual, much lower temperatures, which blunt the snow and stop your skis from moving. Above me the sky is clear, studded with innumerable twinkling stars despite the light of the moon.

Once again, the air is a flurry of ice crystals. They refract the moonlight and a moonbow forms in the sky before me, faint and pallid, as large as a rainbow. The moon is directly opposite; in its light, my head casts a shadow on the snow right in the middle of the arch. What is this phenomenon? Moonbows form when there are liquid drops of water in the air. I guess they must be between the ice crystals? It's not unusual to find liquid drops at −15°C; small drops don't freeze easily and often remain in liquid form at much lower temperatures. The process that causes them to freeze is a crucial part of our research.

▶ Snow crystal preserved in acrylic. In winter, when the temperature is extremely low, we often see these crystals swirling in the air, flashing in our headlamps. The crystal has a diameter of almost half an inch.

But I don't notice any icing, as I would expect in these conditions; at low subzero temperatures, drops floating in the air freeze as soon as they meet a surface, quickly covering everything in a thin layer of ice— our glasses and polar suits ice up very quickly. It's a nightmare for our helicopters, which can end up grounded if these conditions take us by surprise. But here there is no trace of icing. Perhaps there's a very thin, mixed-phase cloud somewhere on the ice before me, a cloud consisting partly of water droplets and partly of ice crystals.

Huge, beautiful snow crystals swirl around me slowly, little stars almost half an inch in size, flat and exceptionally intricate. They flash briefly in the path of my headlamp, reflecting the light into my eyes like tiny mirrors, twinkling magically in the air. Back on the ship, I carefully collect a few icy stars. They are fleeting and fragile and crumble at the slightest touch. Nevertheless, I manage to transfer a few onto slides and preserve them in a special liquid under a glass cover. Over the next few days, the liquid hardens around the ice crystals; the small amount of water in the crystals diffuses as time goes by. What remains is an imprint of the crystal, down to the tiniest detail, white against the hardened, transparent acrylic. I learned this technique from our mechanic, Thomas, who perfected it during his lengthy stay at the Neumayer Station. It allows me to preserve this otherwise ephemeral Arctic masterpiece, a wonderful gift for my loved ones back home.

A storm is forecast for the weekend, with winds of level 7 to 8, and squalls of 9 to 10. With wind speeds of around sixty miles per hour, it'll be a violent storm indeed. I decide to err on the side of caution and cancel the event Marcel has arranged for Sunday—a *Kohltour*, a North German tradition that involves walking through the cold while eating steaming cabbage.

5

STORM IN THE POLAR NIGHT

November 16, 2019: Day 58

SATURDAY MORNING, 7:00 AM. Stefan Schwarze and I stand on the bridge, discussing the day's plans and watching the growing storm. The coffee machine burbles in the background, its aroma announcing the first cup of the day. There's a routine to this morning meeting. As always, it's attended by First Mate Uwe Grundmann (the captain's deputy); Marcel Nicolaus (my deputy); Chief Jens Grafe, who is responsible for the engines and leads the engine crew; radio operator Gerd Frank; and Wulf Miersch, the ship's surgeon, whom we all call Doc. At around 7:20 AM, the on-duty bosun is called up from the working deck to join the meeting. The bosun oversees the deck crew and manages all work taking place on deck on a particular day.

As always, there are a few things to discuss. Did anything unusual happen overnight? Where will the ship's crew be needed today? Will the crane be required, and will it be able to operate in the encroaching storm? Et cetera, et cetera. We've got this ritual down to a fine art so that we can thoroughly coordinate everything happening on board.

Today everything's normal; there have been no unusual incidents and the mood is relaxed. It's set to be a good day.

Suddenly there's a loud bang. A shock reverberates through the ship as it's thrown back and forth and then trembles beneath our feet. There's also a crunching sound coming from the bow. A huge amount of pressure has built up in the ice directly before us, and now the ice has succumbed, breaking apart along the ridge in front of the bow. The ice on our starboard side pushes against and under the ice in front of us. Along the fault line, ice floes rise about ten feet into the air, then sink into the ocean under their own weight. They push beneath the adjacent rising floes and a whole new set of icy mountains emerges. This is what it must look like as our continental plates form new mountain ranges over millions of years. We are watching an extra-quick-motion display of the forces involved in plate tectonics.

The weight of the new mountain range pushes down on vast areas of the ice; water penetrates through the cracks, flooding areas along the new pressure ridge and forming greenish ponds on both sides that shimmer beautifully in our searchlights. An entire landscape has evolved in a matter of minutes with mountain peaks, deep gorges, and large lakes.

We watch on, fascinated. Doc films it on his phone. Soon the ice pressure abates, and the pressing and pushing loses momentum. After twenty minutes, total calm has resumed. But our surroundings have changed. Again. And we need to respond quickly.

We now have a new, taller pressure ridge in front of us. It has closed in on our main power line and threatens to engulf it. Even worse, a whole series of cracks has opened in the ice perpendicular to the new ridge. They're not particularly wide; some are little more than a foot, the widest around three feet. But they are heading directly toward our fortress, and at least one appears to run straight into it. Our tower of strength is damaged!

Along the way, this family of cracks crosses the power line and the road to ROV City (again!) before continuing to the Remote Sensing

Site. It traverses the spine of the camp—which holds our main road and main power line—at Ocean City before disappearing into the fortress at Balloon Town. The cracks branch off in all sorts of directions, particularly around the Remote Sensing Site, and run close to the equipment on the ice, including our most expensive pieces. The whole area appears much less stable now.

A team rushes out to move our main power line further astern, away from the encroaching pressure ridge. Another group heads to the Remote Sensing Site to take a closer look and salvage any vulnerable equipment. Thankfully, the ice dynamics seem to have calmed for the moment, so my colleagues can take their time.

The rest of us prepare the ship for the storm. All nonessential equipment is loaded back onto the ship. We use the crane to lift four of the snowmobiles back on board, safe from the storm. We leave four on the ice in case we need to spring into action before the storm passes. We move our main power line and data cable a little farther away from the new pressure ridge and hope it will be safe.

In the afternoon I put on my skis and inspect the camp again. The wind has now reached level 8 on the Beaufort scale, but it's also incredibly warm. With the storm approaching, the temperature has increased to −8°C (17.6°F). I'm sweating in my polar gear and the strong wind

▶ Our workplace is transformed by a new pressure ridge.

isn't hurting my face; I don't even need total coverage. Not too much danger of frostbite today.

The storm is stirring up the snow into drifts that are only around three feet high, but the snow flurries stop the light of our headlamps from reaching all the way to the ground. I ski into the darkness through a billowing ocean of snow.

It's only just beginning to snow, so I still have a good view above the drift. I reassure myself that if there were any polar bears, I would be able to spot them. The snowfall is setting in, and within an hour, visibility will be so limited that we won't be safe outside. What a difference from the Antarctic, where I traveled in much more violent storms, using guidelines to pull myself along, hand over hand, even though I couldn't see my own feet and could barely see my hands in front of me. But in the Antarctic, the only thing a snowstorm is likely to conceal is a bunch of penguins, and they won't try to eat you!

We've secured the camp as best we can. We've retrieved anything that might fly away or disappear forever into the snowbanks that are already forming. Everyone returns to the ship and we raise the gangway. Bring on the storm!

After shedding some of my many layers of clothing, I make myself a coffee in my cabin, put on some good music—Buena Vista Social Club—and update my diary. We're well prepared, and there's nothing more I can do anyway. I relax (not something I get to do very often).

That might sound a bit strange. There's a storm raging outside, growing stronger by the minute, howling and rattling the ship. We're in the middle of nowhere, fending for ourselves, on an Arctic ice floe that could break at any moment—and I'm enjoying a relaxing evening. Obviously, I check in regularly with the bridge, but even that becomes increasingly pointless. The heavy snowfall is now completely obscuring the view. All we can see is the snow whirling in the squall. And yet there's something very calming about knowing you've done everything you can. In the end, I go to bed early. A good night's sleep is the best way to prepare for whatever happens next.

▶ The *Polarstern* in the gathering storm.

November 17, 2019: Day 59

MY NIGHT IS cut short by the shrill ring of my cabin phone. It's 5:00 AM. The storm continues to buffet the ship, but the snowfall has eased and we now have an unobstructed view of the drama taking place outside.

▶ HOW DOES THE WIND BLOW?

There are many common units of measure for wind speed, requiring frequent conversion. A standard way to express wind speed is in meters per second (m/s), in line with the International System of Units. However, some people may find kilometers per hour (kmh) to be easier to understand. Two traditional systems also remain in common usage. In seafaring and aviation, wind speed is often expressed in knots (kn), or nautical miles per second. The Beaufort scale (Bft) is common in meteorology; this divides the usual wind speeds into twelve categories, from "calm" (0 Bft) to "hurricane" (12 Bft). In the US system, wind speed is often expressed in miles per hour (mph), which is based on (slightly shorter) land miles rather than nautical miles.

▶ **Beaufort Scale**

Strength in Bft	Description	Wind Speed			
		m/s	kmh	kn	mph
0	Calm	0–0.2	0–<1	0–<1	0–1.1
1	Light air	0.3–1.5	1–5	1–3	1.2–4.5
2	Light breeze	1.6–3.3	6–11	4–6	4.6–8.0
3	Gentle breeze	3.4–5.4	12–19	7–10	8.1–12.6
4	Moderate breeze	5.5–7.9	20–28	11–15	12.7–18.3
5	Fresh breeze	8.0–10.7	29–38	16–21	18.4–25.2
6	Strong breeze	10.9–13.8	39–49	22–27	25.3–32.1
7	Near gale	13.9–17.1	50–61	28–33	32.2–39.0
8	Gale	17.2–20.7	62–74	34–40	39.1–47.1
9	Strong gale	20.8–24.4	75–88	41–47	47.2–55.1
10	Storm gale	24.5–28.4	89–102	48–55	55.2–64.3
11	Violent storm	28.5–32.6	103–117	56–63	64.4–73.5
12	Hurricane	From 32.7	From 118	From 64	From 73.6

The storm has done some serious damage. Where yesterday there were cracks in the ice, now there is an expanse of rubble made up of smaller ice floes. They float in a lead (ice-free channel) about a hundred feet wide. This area of open water extends fifty yards in front of our bow, left and right, beyond our range of vision. To starboard, the lead slices through the road and power line to ROV City, continues toward the fortress, cuts across our main road on the spine just behind Ocean City, and then disappears deep into the fortress. At least the field of debris narrows at the entrance to the fortress, where it's largely restricted to a single fissure that's currently around thirty feet wide. In the night, the storm has knocked over the 1,500-pound power distributor at ROV City and the equally heavy main distributor near the ship; they are both on their side with their skids sticking up in the air.

But the most urgent issue is the power lines themselves. The ROV City power line has been torn from its posts and is trapped at an angle in a block of ice, running across the ice like a piece of elastic that's about to snap.

We need to hurry. The cable could tear at any moment and be destroyed. I wake Audun Tholfsen and Hans Honold from the logistics team; while Hans keeps guard from the bridge, Audun and I rush down to the ice to disconnect the ROV City power line—but only once we've armed ourselves with a partially loaded gun, ready to shoot should the need arise.

The storm is still raging: level 9 and above, −20°C (−4°F)—in just a short time, the temperature has fallen by 12°C (almost 22°F). This indicates that the storm's cold front is moving over us. These conditions will freeze your eyelashes together in an instant.

Before we can reach the power distributor, the plug connection rips from the anchor that's securing the cable in the power distributor. The now-currentless power line springs back. Things have worked out well.

At first glance, even the plug connection to the power distributor seems largely undamaged. But the power line that runs from the distributor box toward ROV City disappears in a muddle of ice floes around the lead. It'll take a while to get it out. We lay the loose cable flat on the ice between the debris and the box so that it can't get trapped if the ice moves again. With no power, the warm light of ROV City is extinguished on the other side of the lead.

We look over to the infrastructure along the spine of the camp. As expected, every light on the other side of the lead is extinguished. Met City and the Remote Sensing Site are dead. But the large orange tents in Balloon Town are still shining, a calming glow in the storm and darkness. This means that the Ocean City distributor still has power.

My companion and I dig a snowmobile out of the towering snow-drift, open it up, and free its power train, which has been totally snowed in; after a few minutes' work, it's ready to go. Then we ride to Ocean City. Behind the power distributor we find pure chaos. The three-legged posts—which run power lines and data cables from here to Met City, the Remote Sensing Site, and Balloon Town—are strewn across the ice, and the two thick orange cables keep vanishing under the ice blocks that have been lifted onto the edge of the crack by the ice pressure.

We manage to disconnect the Remote Sensing Site cables and pull them over to us. They have been torn out of the power distributor over at the site, and their ends, when we finally retrieve them from the water, are frayed. The cable to Balloon Town is live and we leave it that way; we just move it away from the crack to secure it. But there's no quick remedy for Met City, which is on the other side of the lead. Just like in ROV City, its cables have vanished beneath the ice chunks and floes. This is a job for a larger team who can take their time.

We return to the ship for the moment and plan a day of cable res-cues. Audun has already attached two kayaks to a flexible catamaran with pieces of wood and cords for just such an occasion. And now it's time to use them. After breakfast, a team of eight pull the kayak-catamaran all the way to the Ocean City crack. They lower it into the water, and two people sail over to cut the Met City power line at one of the connectors. They use the snowmobiles to drag the now flat and loose cable end free of the ice blocks and secure it. They do the same for the ROV City cables. Then the whole team returns to the ship safe and sound.

In the evening, the ice around the crack rises, forming huge bod-ies of water with dozens of smaller floe pieces floating between them. Two of the three lines attaching the ship's bow to the ice floe are now embedded in nothing but tiny blocks of ice splashing cheerfully in the open water. On the portside, the ice gapes at least three hundred feet wide. We now have just one bow line holding us on the floe in winds reaching level 8 on the Beaufort scale. It's under enormous strain and

▶ TRAVELING ACROSS THE ICE

Every footprint and snowmobile track changes the snow cover and underlying ice, potentially interfering with our data. This is why we all stick to our established paths and roads. Expedition members can walk short distances along the paths marked with flags, but we have several other options for getting from A to B:

- Pulks—small sleds that are easy to pull across the ice—can be used to transport devices and equipment.
- Snowmobiles, with a chain drive at the back and skids at the front, are useful for traveling quickly on the ice.
- Nansen sleds are attached to snowmobiles to transport people and equipment. They can also be pulled by hand across short distances. Turned upside down, they make excellent improvised bridges when crossing cracks in the ice.
- Skis are very useful for longer distances.
- Kayaks are used to cross cracks and leads in the winter, and open water and large melt ponds in the summer. When we are transporting larger instruments, it has proved useful to bind two kayaks together to make a catamaran and to make pontoons from hollow plastic blocks.
- The Argo is a small, tracked vehicle used to move loads across the ice floe.
- PistenBullys are large, tracked vehicles that can carry heavy loads and create wide tracks on the ice floe, for example for a landing strip. The ice must be at least three feet thick to bear their weight.
- Dogsleds were common on historic expeditions and are still used at the research station on Spitsbergen, but are not part of MOSAiC. The researchers want to avoid external influences as far as possible, and animal hair could distort their measurements.

could tear or pull out the steel ice anchor at any moment. If it does, there will be nothing holding us here; the ship will be out of control and drift away from the ice edge.

If the worst were to happen, we wouldn't have enough time to start the engines. We confer briefly; just to be safe, the captain calls the engine control room and tells them to start up two of the four main engines, ready to steer the *Polarstern* and maintain control of the ship. If we are forced to give the propellers a burst or employ lateral jet propulsion, the enormous water current from the propellers will finish off our ice floe. What else can we do? If the final line fails, the ship will drift freely in the water and we'll need to be able to maneuver. After several weeks without the need for engines, the *Polarstern* once again vibrates with the thumping of the pistons. Is this the end of our research camp?

November 18, 2019: Day 60

THE MORNING STARTS with a surprise: the ice pressure increased again during the night and pushed everything back together. It's some sort of miracle—dozens of tiny pieces have all returned to their old positions. The floe has reassembled itself like a puzzle; MOSAiC really was the perfect name for this expedition! We allow the *Polarstern*'s main engines to resume their hibernation. The ship is secure in the ice, so we won't be needing them.

We go outside to inspect the camp and figure out which sections can be reached safely. We're still in the storm, with fairly high wind speeds of forty knots (around 46 mph). Everywhere we look, pieces of floe debris press together, yawning cracks between them up to ten feet wide filled with ground-up, freshly frozen slush. Many cracks have been covered with drift snow by the wind, rendering them invisible—and treacherous for anyone trying to make their way across the ice. If you failed to spot a crack and placed your weight on it, you would go straight through to the icy water. If there's any doubt, we always use a stick to check whether the ground is safe.

▶ Improvising bridges using upturned Nansen sleds and wooden pallets. This is often the only way to cross the many cracks in the ice and access the research cities.

We fashion a series of makeshift bridges to negotiate this icy labyrinth. We push Nansen sleds over the cracks and turn them over; they make excellent mobile platforms as the ice floes move and shift beneath our feet.

We've been lucky amid all this mayhem. The installations seem to be generally okay. Nothing has sunk into the ocean or seems to be in serious jeopardy. The ninety-eight-foot-high meteorological mast has suffered in the storm and weaves a crooked line into the black sky, but the guy ropes are keeping it upright and it looks like it can be repaired.

The large hangar in Balloon Town is slightly askew and threatens to crush Miss Piggy, our tethered balloon. The hangar is held up by hoses inflated via extreme pressure. Whenever squalls hit the tent, the hose relief valves activate and release air with a hiss. Unfortunately, the automatic control system for the compressors—which are supposed to replenish the air in the hoses—switched itself off during the power failure and didn't start up again. We close the relief valves, get the compressors working, and straighten the hanger up. It should be able to weather the rest of the storm.

▶ The Balloon Town hangar, upright and stable after the storm.

The most urgent work is complete, and we return to the ship. The storm slowly abates as afternoon approaches, and we divvy up the remaining tasks: right the power network posts knocked over by the storm and replace them where broken, restore the entire power network, re-create the snow-covered paths with shovels and pickaxes, mark dangerous spots where the ice floe is fractured, and so on. Teams fan out across the ice, ready for a busy afternoon, and by evening we're back in business!

November 19, 2019: Day 61

MIDNIGHT, DEEPLY ASLEEP. A call from the bridge: the mast in Met City is down!

From the bridge, I can see that the entire ice field southwest of the crack zone has shifted fifty to a hundred yards to the portside, taking Met City and ROV City with it.

The crack zone itself has closed up again, more or less, but an unremarkable narrow crack has formed diagonal to this zone. It runs right between the base of the meteorological mast and its anchor, where two guy wires have been stuck into the ground. This was all it took to shift the tower sideways on its base and pull its feet out from under it. It has collapsed and is now lying over the crack, a twisted heap of rubble. There's nothing we can do right now.

Once day has broken—or what passes for day in the polar night—we try to find a way to Met City through the chaos of cracks and ridges. We salvage the tower ruins and the instrument that sits at its peak. It's not looking good; the ultrasonic anemometer, which measures turbulence in the air, is bent out of shape and probably irreparable. But we do have other instruments that perform the same function for the mobile thirty-six-foot tower, including a replacement device that isn't currently in use and might be suitable once the large mast has been rebuilt.

As the day goes on, the crack zone shifts several times, moving farther to port. Every time, we hear a grinding, grating sound in the darkness as the floes press along one another. Where friction builds, little pieces of ice debris form new, narrow ridges. After around fifteen minutes, the movement stops as abruptly as it started.

▶ Near to Ocean City, a sixteen-foot-high pressure ridge forms out of nothing. Just two hours before, this was all flat ice. It threatens to engulf the station, which has to be relocated.

November 20, 2019: Day 62

TONIGHT, THE BRIDGE summons me from my bed at 11:30 PM. The area to the south of the crack zone has started moving and is gliding rapidly to the east. We watch the whole thing from the bridge in the searchlights. One by one, as if by magic, ROV City, the Remote Sensing Site, and Met City drift slowly past our bow. The watch officer is known for his black humor, and he plays "Time to Say Goodbye" over the bridge's loudspeakers as the cities disappear into the darkness on the portside. The world around us is moving, and all we can do is watch, powerless.

As the ice floes grind together, more towering ridges emerge in the crack zone. On our side of the crack, a sixteen-foot-high mountain forms near Ocean City out of what was previously flat ice, threatening to swallow the research station. On the other side, a ridge moves ever closer to the remote sensing instruments; damage is imminent.

At around 1:00 AM, the movement slowly stops. The layout of our research camp has changed completely. The parts beyond the shear zone are now six hundred yards farther east; once again, ROV City is farthest away. I go back to bed; we can't do anything right now.

I don't sleep for long. After less than an hour, the phone rings again. Three polar bears are approaching the research camp from the north and have triggered two flares on the trip wire. They leaped back from our installations on the ice, startled, but are now walking placidly around the back of the ship, maintaining a distance of around five hundred yards and looking at us curiously. It's a mother with two half-grown cubs. One of the cubs explores everything he sees, running to every flagpole and device, always bringing up the rear. Visibly annoyed, the mother keeps turning around to badger him, and then he dashes to catch up. The other cub stays by his mother like a good little boy. Clearly, the two have already developed individual personalities.

The three bears round the ship via the portside until they come to a crack. It takes them a while to find a way across. Here and there the mother climbs onto a block of ice to get a better view. They obviously don't want to get their feet wet! Eventually she finds a way and guides her young over the crack, safe and dry. They continue around the ship, the curious young bear examines all our equipment on the edge of the path, then the three disappear into the darkness on the front starboard side. We don't see them again.

The next morning, the situation in Ocean City grows increasingly perilous. We decide to evacuate it when a new crack opens right beneath the research hut with the ice hole for the water probe. We work late into the night, digging the tent and its precious contents out of the crusty snow and freeing them from the ice. The hut's foundations consist of a floatable platform with individual hollow plastic blocks stuck inside one another. Eventually we get the structure and its foundations out of the ice and use the Argo, our tracked vehicle, to pull them to the secure logistics area next to the ship. Ocean City is history.

The mood is gloomy. It took weeks of hard work to set up our research camp, and now it has suffered significant damage, its future uncertain.

Safety on the Ice

ON AN EXPEDITION like this, unexpected developments crop up every day. We have to respond and adjust our plans accordingly—or, as we are currently finding, abandon them altogether. Quick decisions must be made every day and every hour with safety as our top priority.

It's often necessary to weigh things up carefully. We've come here to perform research that is as good and comprehensive as possible. But we can only do that by ensuring that all expedition members are safe at all times, despite hindrances such as darkness, cold, polar bears, and precarious ice that constantly moves, cracks, and forms ridges. At the same time, our setup has to allow us to conduct an extremely diverse scientific program that we have been preparing for years.

Every time a team goes onto the ice, they have to follow the equipment rules. We wear special polar suits that don't just keep us warm but will also keep us buoyant if we fall through the ice. The wearer floats on the water's surface like a cork, significantly reducing the risk of getting trapped under the ice. We also carry lifelines—floatable throw lines stored in bags—on carabiners on our belts, always at the ready. If someone does fall through the ice, these lines are an indispensable form of rapid assistance. In addition, each team member carries ice picks in their breast pocket, little metal pickaxes that we can use to pull ourselves out of the water and back onto the ice. We practiced this before the expedition. In an emergency, you only have a few minutes before your arms become stiff and immobile in the chilly Arctic water.

Every group that leaves our camp's central observatory carries equipment that will help us to find them should they require rescue: the obligatory radio device (a satellite phone with extra battery), an in-reach device (which determines GPS positions and can transmit to the bridge and send messages), and a sort of digital beacon, which sounds a shrill alarm when activated and appears on the ship's radar to determine its exact location.

The farther a team moves from the ship, the more equipment they take. The emergency backpacks for distances of up to three nautical miles (around 3.4 miles) contain a first aid kit, changes of clothing, and a bivouac sack.

We have a big emergency pack for longer excursions, for example if a team flies by helicopter to check and maintain one of our distributed network stations, which are far beyond the reach of our radios. The helicopter returns to the ship while they do their work, leaving the team alone. They maintain contact with the bridge via satellite phone. Teams have set check-in times; if we don't hear from them, we send a rescue party. If the group can't return to the ship as planned—because the weather has changed, for example, or there's a problem with the helicopters—the emergency pack has everything they need to survive a few days in the Arctic, from a tent and sleeping bags to folding spades, lighters, food, a stove to prepare drinking water, and first aid supplies.

▶ CONNECTING TO THE REST OF THE WORLD

These days, we're accustomed to top-quality, worldwide communication; you can even transmit from the farthest reaches of the Amazon, live and in high definition. The little areas around the North and South Pole are the last places on Earth to be cut off from the global telecommunications network. This network uses geostationary satellites, which have a fixed position above the Earth's surface and must be 22,000 miles above the equator for reasons of orbital mechanics. You can contact them from almost anywhere on the planet—just not from the central Arctic or Antarctic, where the satellites are positioned directly on or just below the horizon. These regions are limited to rudimentary telecommunications via a network of dozens of satellites that orbit just a few thousand miles above the Earth's surface, some of them passing regularly over the poles. Depending on where these satellites are currently situated, you can radio them, and they will pass the signal through the network until it can be sent to a ground receiving station in an inhabited part of the world.

This network of satellites allows us to make telephone calls, although the quality is often limited and the delays can be long. We can sometimes transmit data packages too, although this is also rudimentary and sluggish. Calls get cut short when the satellites alter their configuration, and the data rates are nowhere near good enough for video images—services like Skype are out of the question. Short text messages can be sent via WhatsApp (often with delays of several hours) and many team members use this method to stay in touch with their loved ones back home. It isn't usually possible to send photos.

During MOSAiC, we use two Kepler satellites that are circling the Earth at a low orbit and regularly pass over the polar regions. During these short contact periods, larger data packages can be transmitted to these satellites via a special antenna system that uses a parabolic reflector to follow the satellites' orbit across the sky. The satellites save the data packages and transmit them to a land station as they pass.

The Arctic doesn't forgive carelessness. It isn't always easy to keep your bearings on the sea ice; dense fog can descend in a matter of minutes, there are very few landmarks, and the landscape is largely uniform. It's dark in winter and there's a danger of whiteout in summer, when the white sky blends with the ice and contours vanish. You can't even rely on your own footprints; they can quickly be covered by snowdrifts, which impair visibility and can render navigation impossible.

To travel the Arctic safely, you have to learn to read the ice. After a while, you start to recognize the subtle changes on the surface, signs of cracks hidden beneath the snow or deep drifts in which you can quickly sink up to your navel and struggle to get back out. But no matter how much experience you have, you never stop using metal rods to test the ice before you take a step.

Despite all the safety precautions, nobody can help you if they don't know where you are. Your companion on the ice isn't just your colleague—they could save your life. They will be the one to throw you a lifeline if you crash through the ice, the one to tell you that your face is showing signs of frostbite.

Just as in Nansen's day, if you want to stay safe, you have to be able to rely on one another.

November 22, 2019: Day 64

ONE DIFFICULT DAY is followed by another, and another. The expedition is a marathon, and if you don't achieve something one day, maybe the next will be better.

At 8:30 AM on the dot, Hans Honold's genial Bavarian tones greet us from all radio devices on the ice. Hans is a member of the logistics team, and his morning announcements have evolved into a top-quality show: Radio North Pole, definitely the world's most northerly radio station! He plays something different every morning to get the day started. Today it's the Rolling Stones. His brief broadcast lifts our spirits and fires us up. Little things like this are worth their weight in gold when times are tough and we're down in the dumps. Hans has good instincts and keeps us all going. We really need it right now.

The damage to our research camp has destroyed weeks of work, but there's no point dwelling on it.

Yesterday we started to rebuild the camp, and everyone's giving their all. As the day went on, Met City began emergency operations using a power generator transported over by a team. The Remote Sensing Site has been secured, its shelter freed from the ice and relocated; we moved some of the expensive instruments to safer areas of the ice. A detailed survey showed that very little equipment has been lost forever or buried under piles of weighty ice blocks. A crack formed in the (as yet intact) part of the spine between the Remote Sensing Site and Met City and continued to open as the day went on, so one group dismantled the power lines along the spine and moved them over to Met City's side. We decide to wait until morning, see how the ice has moved, and then start planning the camp's future layout. One thing we do achieve today is to reestablish Ocean City near its old spot, but on safer ice. We can still reach it easily on foot and snowmobile from the ship.

In the afternoon, we examine the Eco team's dark site, far from the lights of the *Polarstern*, where the biologists are studying light-sensitive microorganisms. We name it Mordor, after Frodo's treacherous destination in *The Lord of the Rings*, which was shrouded in constant darkness. When we arrive, we're relieved to see that everything's intact; there's just a small crack across the path. We mark it with red flags. We can cross the crack on snowmobiles, provided we don't overlook it and snag one of the skids. If you ride over it too fast and at the wrong angle, you'll fall head over heels and might pick up a serious injury.

Behind the dark site is a vast plain of new ice, which covered the freshly formed lead. As the expedition continues, this will prove to be one of our most valuable ice surfaces; we will be able to study the young ice in detail for the rest of the year and beyond, all the way into next summer, and watch the birth of the ice—a scientific windfall!

In front of it we can see beautiful ice sculptures in the dark, ossified in the polar night after all the movements of the last few days. Far in the distance, the *Polarstern* is a dot on the horizon; in the endless black of the night, with nothing but a tiny spot of light from your headlamp, you can't help but feel small.

▶ When you work on the ice, your world shrinks to the light cast by your head-lamp in the black night and vast expanse.

November 26, 2019: Day 68

THE REMOTE SENSING SITE is moving house today. ROV City has already been rebuilt nearer to the other sites. The research camp is slowly becoming more compact, after immense shifts in the ice spread it all out and made it difficult to protect ourselves from polar bears. Now we can connect everything to the new power network; step by step, the sites running on emergency generators are returning to full operations. We're getting there!

In the afternoon we perform a rescue exercise on the ice. All expedition members completed training before departure—how to extinguish a fire on board, how to rescue someone who's fallen through the ice, and how to free yourself from a helicopter sinking into the sea (using the emergency breathing apparatus we always carry in our breast pockets when flying). Now we practice the procedures for accidents on the ice—the role of the bridge (which coordinates the emergency response), first aid measures on the ice, and the recovery team's transfer of the injured to the *Polarstern* hospital, where Doc will be waiting to care for them. It's an excellent performance: thirty-five minutes from raising the alarm to operating on the "injured."

This evening, again, vast compressions make the ice tremble. The ship judders back and forth. A tall ridge develops near Balloon Town and Ocean City, the location of our path across the gaping shear zone. We decide to scout for a new crossing, this time nearer the ship.

▶ WHAT IS THE WORLD CLIMATE CONFERENCE?

The topic of climate change may only have gained public momentum in the last few years, but it has been on scientists' radar for much longer. The first World Climate Conference took place in 1979 under the auspices of the United Nations. Since 1992, countries and non-governmental organizations have met regularly to develop a common strategy against climate change resulting from human activities. In 1997, the Kyoto Protocol was signed in Japan, setting the first joint and binding targets for reducing greenhouse gas emissions. In its successor, the Paris Agreement, the UN member states committed themselves to limiting global warming to below 2°C (3.6°F) compared to preindustrial levels. However, the plans that these countries have put forward so far are yet to achieve this goal. And even if the 2°C limit is not exceeded, our planet faces irreversible consequences.

November 27, 2019: Day 69

THE *KAPITAN DRANITSYN*, a Russian icebreaker, sets off from Tromsø today, carrying the expedition members for the second phase. It will take them weeks to reach our position deep in the ice. They will relieve us from sometime in mid-December onward. Some members are staying on board for the second phase; others will be returning later on. I intend to rejoin the ship in mid-March and remain on board until the end of the expedition. In the meantime, my experienced colleagues Christian Haas and Torsten Kanzow will each take charge for a few weeks; they both said yes immediately when I offered them this crucial role. Torsten Kanzow will end up staying on board much longer when MOSAiC runs into difficulties in March.

A few days ago, the captain and I investigated a possible route for the final approach of the *Dranitsyn*. We decided to proceed with the plan we'd had in mind for some time: The *Dranitsyn* is to approach from behind on the portside, more or less directly from the north, where we sailed into the ice floe all that time ago. She will then stop with her bow near our portside stern. This is the best position for swapping tons of freight with the ship's cranes. Personnel will move between the ships on the ice, via the logistics area, and when the *Dranitsyn* departs, she will sail backward along her own approach route—the shape of her hull allows her to break through ice efficiently while traveling backward. This is a good plan that should minimize damage to our ice floe and the research camp. The *Dranitsyn* has already been delayed; just hours after leaving Tromsø, she was forced to shelter in the fjord from a violent storm in the Barents Sea and still isn't able to leave. Her foreship is fitted with refrigerated containers for our operations, making her susceptible to the force of the waves, and those can't be avoided on the open sea. She will be stuck in the fjord for several days.

November 28, 2019: Day 70

IT'S MY BIRTHDAY! As soon as I wake up, I open all the presents my family packed for me back home. They include our traditional Christmas stollen cake, so I don't have to miss out if I end up spending Christmas deep in the Arctic. Later, I treat myself to a long skiing trip out into the night, two and a half hours of rapture in the landscape and extraordinary ambience of this dark, frozen world.

This truly is a sublime place to spend your birthday. In the evening we celebrate with mulled wine and grog at our ice bar near the ship. It's almost −30°C (−22°F) so we drink our wine fast; otherwise it'll freeze in our cups. The general rule is that the first mouthful is warm, the second cold, and the third frozen. It doesn't harm the mood, though; we enjoy a wonderful party surrounded by ice beneath a sky of glittering stars.

November 30, 2019: Day 72

SINCE YESTERDAY, PRACTICALLY every station has been connected to the power network, and the camp is up and running! We hear occasional rumbling in the ice, but there have been no more major shifts. The temperature stays between −25°C and −30°C (−13°F to −22°F), and every crack quickly freezes solid. Could this be the start of a longer period of stability?

December 1, 2019: Day 73

IT'S THE FIRST SUNDAY of Advent! The crew have decorated the entire ship; wherever you look, something dangles and glitters. Finally, we can catch up on the *Kohltour* we've postponed several times due to polar bears, the storm, and urgent work. With a music system in a backpack and elaborate drinking cups fashioned from ice by Marcel Nicolaus, we stroll across our ice floe, passing some of our cities. We pause at every one, and when we reach Met City, some people start to dance. Then we eat the traditional green cabbage; our chef has rustled some up from his stores and added savoy cabbage (which we hardly notice) so that there's enough to go around.

I finish this wonderful day by sharing whiskey and cigars with the captain. Antje Boetius (who's always looking out for us) gave us a little box of good cigars before we departed. I'm sure she wishes she were here with us now.

December 2, 2019: Day 74

TIME TO GO exploring again. Thomas Sterbenz, our vehicle fleet expert, joins me as I travel far beyond the boundaries of our central observatory. We want to find a place for aircraft to land. At the moment, we're focusing on creating a landing strip for DHC-6 Twin Otters—small planes—so that we have an emergency evacuation option, an invaluable part of our safety precautions. Previously we wouldn't have dreamed of landing anything on the ice because it wasn't thick enough. But now it might be.

We find a flat field and use the drill to measure the ice thickness along what could be our landing strip. We mark out over half a mile with our flags, with the potential for later expansion. The ice is over

▶ People working on the ice can end up frozen beyond recognition. Eyelashes grow heavy with ice and freeze together. When the temperature drops even further, you can't do anything without full face masks and protective eyewear.

three feet thick for around six hundred yards, plenty for a Twin Otter. The surface here is so even that it would require little preparation. Should we need to evacuate, we could set up the runway pretty quickly with beacons. This is a huge boost to our safety; we've long since drifted out of range of the Russian long-distance helicopters. In the months before the expedition, we set up fuel depots on the most northerly islands off the Siberian coast so that the helicopters could assist with evacuations in the initial expedition phase. Our emergency operational plans, which involve two helicopters (so that one can rescue the other in the event of an accident) are sitting in a drawer. But the drift has taken us out of their range.

It's beautiful out here, so far from the ship in the absolute darkness.

December 3, 2019: Day 75

THE DAY GETS OFF to a bad start. The crack has opened again and is too wide to cross on snowmobiles. Once again, we can only reach ROV City, the Remote Sensing Site, and Met City on foot by constructing makeshift bridges over the cracking and crunching ice.

The wind freshens in the morning, and at around 9:00 AM it begins to snow. The wind reaches gale force again and snowdrifts set in. At 10:15 AM, I decide to vacate the stations separated from the main floe— ROV City, the Remote Sensing Site, and Met City. Fifteen minutes later, I recall the teams from Ocean City and Balloon Town, the fishing group, and the people repairing the trip wire. By 11:00 AM everyone is back on board—but someone's missing! We radio everyone, and eventually our colleague is located on the ship; he forgot to update the logbook when he came back. Now we can raise the gangway.

In the afternoon, the crack pushes forcefully toward ROV City, and a new icy mountain range begins to emerge along the previous crack, as the pieces of ice are pushed sideways and pressed together. The ice blocks quickly pile up until they're many feet high, burying the Nansen sleds that were parked there to bridge the crack.

The power line and a box of important equipment also end up in the mass of ice. We quickly lower the gangway, and I rush over with a small team. Every time the movement slows, we seize our chance to rescue what we can from between the mountains of ice blocks. Climbing over the jumble of ice, we manage to salvage the box and extricate the Nansen sleds from the ridge; they're only partially submerged beneath little pieces of ice. Unfortunately, the power line is stuck fast under tons of ice and will stay there for the next few weeks. We secure both sides and lay out enough cable so that, should the ice shift again, the cable can increase in tension without tearing.

At times, the snowdrift reduces visibility to just a few yards on the ice ridge. Ice and snow crystals whirl around us and hit our eyes like bullets. Without ski goggles we're useless. The snow gets into the fur collars of our hoods and buries our faces as the storm rages around us; it's all we can do to recognize the outlines of the ice blocks right in front of us. The whole ground is moving. Eventually we return to the ship, our mission successful.

December 4, 2019: Day 76

THE STORM PERSISTS, and our time on the ice is brief; after that, the driving snow means visibility is just too poor. At least the ice is stable again: no shifting, no channels. We receive a message letting us know that the *Dranitsyn* left the fjord this morning and is on its way. When will we see our homes again? Might we be there for Christmas?

6

CHRISTMAS
IN THE ICE

December 5, 2019: Day 77

TODAY THE *DRANITSYN* reached the ice edge to the east of Spitsbergen. She's progressing through the young ice (about sixteen to twenty-four inches) at eight to nine knots. That's a good speed!

It's still very stormy here. Our work on the ice is limited. I'm beaming in live to the COP25 World Climate Conference in Madrid. From the very top of the planet, I'll be taking part in an event that attracts a global audience. The signatory states of the UN climate accord will be discussing how they intend to achieve their climate goals—in particular, reducing greenhouse gas emissions—to curb global warming.

This year, the conference slogan is "Time for Action." And that time is now. I make this very clear in my concluding statement direct from the epicenter of climate change:

> *We are locked into massive ice in fierce cold, maybe a place where climate warming is not obvious to everybody. But the change is everywhere. The thickness of the ice is only half of what it was at the times when Fridtjof Nansen was on a similar expedition 125*

years ago and the temperatures we measure are five to ten degrees higher than those he observed. We are here to understand how these massive changes affect the climate system and to explore what that means for the stability of the Arctic climate system.

Our mission is to provide the robust scientific basis for the important decisions our societies have to take to shape our future.

It is your duty to acknowledge the overwhelming scientific evidence and to come to the conclusions which are needed, to agree on urgent actions to mitigate further climate change.

If we do not reduce our net greenhouse gas emissions massively, if we fail to reduce them to near zero by the middle of the century, irreversible change of the Arctic will have major effects on the climate in the rest of the world.

If we do not succeed with implementing these actions, our generation will be the last generation which sees a year-round ice cover of the Arctic Ocean.

If we fail, the Arctic will be a different world for future generations, a world with an open ocean where there is eternal ice today, a world in which the North Pole can be reached with a normal sailing boat, where the polar bear is extinct and where the weather patterns driven by the warm and open Arctic result in increasing weather extremes all over the northern hemisphere.

The societies of our world are free to decide on their actions. The pressures from many sides are high. And climate change is only one of the challenges our societies face. But it is one of the most important and most pressing challenges.

In our democracies the fantastic freedom for everybody to decide about the future comes along with the responsibility for everybody to protect the interests of future generations. Our societies need to understand the consequences of their decisions today and they have to take responsibility for these decisions. It is our mission to provide the scientific facts so that these decisions can be made in full understanding of these consequences.

And you, the leaders of the countries of the world, have the duty to come to the right conclusions today. In the interest of the future of our planet.

December 6, 2019: Day 78

IT'S SAINT NICHOLAS DAY! In the mess (the room where we eat), there's a little bag waiting for each of us containing a chocolate Santa and chocolate candies. We ran out of chocolate weeks ago, so we're all delighted.

The wind subsides as the day goes on, and by the evening it's almost calm. It takes the gibbous moon hours to move along the horizon, glowing orange. Around the North Pole, the sun and moon move parallel to the horizon; now that it's winter, the sun is so far below the horizon that we can't see it. The moon rises and sets as it cycles through its phases, rather than in a daily rhythm. From around this phase—the gibbous moon—the moon moves above the horizon, becoming visible. It will spiral higher and higher until the full moon, after which it wanes, sinking back down to the horizon before disappearing altogether.

▶ In the polar night of early December, the *Polarstern* is frozen into the ice as she drifts to the North Pole.

Mirages distort the orange gibbous moon on the horizon. Sometimes it's broken into three orange lines, sometimes it takes on the shape of a glowing mushroom cloud. A beautiful spectacle that lasts for hours.

We finally have the conditions we need to ramp up our research. We're keen to collect comprehensive data, so our operations run twenty-four hours a day; various measurement programs run through the night, and even Miss Piggy, our tethered balloon, doesn't sleep.

In the evening we receive a message from the *Dranitsyn*. She's currently east of Franz Josef Land on the northern border of the Barents Sea, still 420 nautical miles away. From there, she will pivot east to bypass the thicker ice to the north.

The satellite images show a channel running from the Arctic Cape—the northernmost point of Severnaya Zemlya—to our current position in the far north. The *Dranitsyn* is now heading straight for this channel. If she gets there quickly and can follow the channel, she might arrive earlier than expected.

I've texted my family to say there's a remote chance I'll be home for Christmas. My younger son, Philipp, nine years old at the time, has one word in bold at the top of his Christmas list: PAPA!

▶ The *Polarstern* in the ice.

December 7, 2019: Day 79

THE ICE HAS BEEN STABLE for a few days now. All cities are solid and fully operational, connected to the power, data cables, and our roads. We prepare to rebuild the wrecked ninety-eight-foot mast in Met City.

The *Dranitsyn* reaches the Arctic Cape in the evening, and is now just 350 nautical miles away. I run some calculations, based on my assumptions about routes, ice conditions, and how much time we'll need to transfer fuel and freight between the ships. If the conditions are perfect and the ice is thin, it'll be touch and go whether we make it back for Christmas. To be on the safe side, I book a transferable flight from Tromsø to Berlin on Christmas Eve. I don't want to reach land and find that all the flights are full!

December 8, 2019: Day 80

THE MET MAST is back up! But it's only seventy-five feet tall, not ninety-eight. Some parts of the structure were too twisted to be rebuilt. It won't have much effect on the measurement values, though.

And it's the *Polarstern*'s thirty-seventh birthday! In the evening we toast the old girl—and the success of the first phase—with a sherry in the captain's quarters.

December 9, 2019: Day 81

THE BIG CLEANING DAY. We empty and scrub the labs, corridors, and storage rooms. We want to make sure everything's in order for the next team.

And we prepare for the *Dranitsyn*'s arrival. We flag the approach to our ice floe and mark it with buoys. We need to guide the massive ship through the ice without risking damage to the floe. It's a tricky maneuver, but now we're ready.

For days now we've been heading north without deviation. Finally! This is the first time the drift has carried us directly north.

We have spent years investigating the sea ice drift, performing countless studies, calculating statistical scenarios to determine the best region for the drift. But since we attached ourselves to the ice floe, we've found that the wind frequently blows from the east and pushes us largely to the west, rather than bringing us closer to the North Pole.

As expected, our drift position data shows a series of erratic side-steps, detours, and loops. The Transpolar Drift doesn't move uniformly through the Arctic Ocean; it's chaotic, drifting to and fro, while still gradually guiding the ice from the Siberian Arctic across the polar region and into the Atlantic. But as it does so, the drift can go in any direction. Every day is different. We often drift in small, regular circles along a line determined largely by the wind. Pressure and openings in the ice can also follow this cyclical motion. Nansen observed these cycles too, and suspected that they were affected by the tides. With our measurements today, we can distinguish between motions forced by the tides and the internal oscillations in the ice from eddies in the water column below. At these latitudes, both have a similar frequency and can easily be confused. But our measurements confirm Nansen's hypothesis. The motion of the sun and moon is responsible for the little circles we describe as we continue on our way.

December 10, 2019: Day 82

WE'VE FOUND FRESH polar bear tracks in the camp. We search the area thoroughly but can't see any bears. They've probably moved on by now.

The *Dranitsyn* is making slow progress; the ice is causing difficulties. She'll probably still reach us sometime tomorrow, but first she has to navigate the network of stations we've built on the ice some distance from our floe, without colliding with a single one. Some of these stations only send position data every few hours when they communicate with us via satellites. In the meantime, everything has drifted farther, and data isn't much use if it's several hours old. So I program a tool that corrects the station position data every minute using the drift data from the *Polarstern*. We also develop an approach route based on the stations' bearings to the *Polarstern*, allowing the *Dranitsyn* to navigate safely through the network.

In the evening I begin to pack. The first expedition phase is drawing to a close. It's been many years in the planning, and I can still recall the moment when it became my life's mission.

A Fascinating Idea Becomes Reality

WHEN MY CELL PHONE rang, I was in the southern hemisphere, sipping a coffee on a volcanic slope and gazing at a blue subtropical sea. I was on Réunion, a remote and beautiful island in the Indian Ocean (and a French overseas department) that is pleasantly warm all year round and surrounded by vibrant coral reefs. In summer 2015, the island hosted a networking conference for EU researchers, and I was invited to speak. I have always viewed research as international and have assembled many large and successful multinational teams.

The phone call was from Klaus Dethloff. He asked whether I'd like to take the lead in a project that he'd been pursuing and promoting doggedly for years, an idea we'd discussed many times: MOSAiC.

I knew my answer before Klaus had even finished the question. There was a tremendous power to this idea, an idea that would thrill any polar researcher. I knew I was going to say yes. Still, I asked for a day to think it over. I'm a rational person, and my head needs time to catch up with my heart. I accepted by email before the twenty-four hours were up.

We honed the expedition plans in collaboration with Matthew Shupe, an atmospheric physicist from the University of Colorado Boulder and the US National Oceanic and Atmospheric Administration. Uwe Nixdorf, vice director of the AWI and our head logistician, had already calculated that even if we only moved with the ice drift, we would need fifteen tons of fuel per day to generate heat and power and operate the ship safely. The *Polarstern* has a bunker capacity of three thousand tons, so a yearlong expedition wouldn't be possible without additional supplies. But it was enough for up to six months of self-sufficiency—long enough to survive the winter and spring, when the Arctic ice is so thick that (we thought) no other icebreakers would be able to reach us. However, we would be reliant on deliveries before and after this phase.

The expedition was feasible, but we couldn't do it alone. We needed more icebreakers, and that meant international partners. We needed to generate global interest and support for our project, to make it a pioneering achievement shared by the leading nations in polar research and their respective icebreakers.

And we needed funding. Some quick math showed that even tens of millions of dollars wouldn't be enough. The expedition could only happen if the funding reached nine figures. There was no way that any one country could rustle up that much money. Our task now was to get organizations and scientists on board from around the world, to win over ministries and potential funding agencies in various countries. I began a whirlwind tour of potential logistical, scientific, and financial partners. I didn't get to see my family much.

But it was worth it. The idea met with universal enthusiasm. Before I even began to speak, many of my audiences were aware of just how urgent this research was. Eventually we found a realistic way to collect the data we needed so badly from the Arctic—a massive feat of international collaboration that would become a defining element of MOSAiC. Ultimately, funding from twenty and participants from at least thirty-seven nations supported the expedition.

I was delighted to find that, despite their sometimes contrary geopolitical interests in the Arctic, our potential partner countries were able to collaborate without animosity. They all recognized the necessity of better understanding the Arctic climate system. Russia said yes, the United States said yes, China said yes. The momentum just kept on building. Slowly but surely, a dream became extremely tangible.

We also had the full support and active assistance of Karin Lochte, then director of the Alfred Wegener Institute. Her express support for the project encouraged our partners to trust the plans. Karin's calm and persistent approach to helping drive MOSAiC forward was instrumental. By this point, more countries were getting in touch to volunteer assistance and finance. This wasn't my first major research project with a significant budget, but I'd never experienced anything like it.

This was a complex project with an enormous scope, so we needed the right management structures and controlling and steering groups. We had to plan each individual element in detail (and there were many individual elements), make sure they interacted efficiently, and stay within the overall project budget. And we had to do it all while involving more than eighty partner institutions from twenty countries, with no template whatsoever, while remembering that, ultimately, we would be in nature's hands. It took years to figure it all out.

In September 2019, just a week before our planned departure, an enormous logistical operation was underway in Tromsø harbor. A team had been unpacking dozens of freight containers for almost a week. The various scientific groups had sent their containers to our agent in Tromsø, who had passed everything on to the harbor. Fitting it all onto the ship was going to be tricky. Plus, different types of freight had different requirements: some containers were warm, others cold, some needed power connections to maintain air-conditioned temperatures. Then there were the lab containers and measuring instruments that had to be set up in certain ways: an unobstructed view of the sky for the remote sensing instruments, proximity to unpolluted outside air for the atmospheric chemistry labs, and so on. We spent ages pondering the plans, figuring out what would go where and how to make it all work. And now our plans got a reality check.

We ended up with huge piles of unregistered expedition goods that somehow had to be squeezed onto the ship. And just when it seemed that we'd accommodated everything, another container would arrive unannounced. By some miracle—courtesy of our logistics team and the *Polarstern* cargo officers—everything made it on board. The ship had never been so full.

But we were still missing some key equipment. The ammunition for the guns arrived just in time for our polar bear guards. The cartridges for our trip wire flares got stuck in customs; without them, we'd have no protective fence. Even worse, the night-vision devices we needed to spot bears in the polar night were still on their way. They were developed for the military and their export licenses hadn't been clarified. Then the autoclave in the *Polarstern* hospital stopped working, and we had to fly in a technician just to repair it—without an autoclave, we wouldn't be able to sterilize surgical tools.

The problems just kept on coming in those last days in Tromsø, often only minutes apart. We had no idea whether we'd be able to depart as planned. In the end, I checked off every item on my list. The voyage would begin on time.

Now, on board the *Polarstern* in the middle of the Arctic, all that hustle and bustle seems like a lifetime ago. Outside my cabin window, an entire research camp has been constructed on the ice. But we wouldn't be here without those intensive preparations.

DECEMBER 11, 2019: DAY 83

IN THE MORNING, a tiny dot of light appears briefly on the horizon—the *Dranitsyn*! We know this must be an illusion; she's still much too far away. What we're seeing is a mirage of a ship that, in reality, is still far beyond the horizon. It's as though someone has quickly pulled back the curtain to give us a glimpse of the future.

We're all excited for the ship to arrive. It feels like Christmas morning before the presents are opened; we can hardly wait to get started and work hastily to complete our final tasks. There's so much to do before we can hand over to the next team!

First, I fly to the *Dranitsyn* by helicopter. I'm joined by Felix Lauber, the cargo officer, who will stay on board the *Dranitsyn* to guide her through the station network around the main floe. His counterpart on the *Dranitsyn* will travel back with me to the *Polarstern*.

▶ RESCUE FROM THE ICE

The Arctic is a dangerous place. When you get this far north, providing rapid assistance in an emergency isn't just a matter of honor, but a survival strategy. Some rescue missions capture the global imagination, and there's no way of knowing whether the outcome will be spectacular, tragic, or successful. The missing Franklin expedition in 1845 triggered a spate of rescue voyages and heralded a new era of polar travel. Each mission revealed new pieces of the puzzle, getting closer to the expedition's fate. But the remains of the ships themselves were not discovered until 2014 (*Erebus*) and 2016 (*Terror*), almost 170 years after they vanished.

Some budding rescuers have ended up in sticky situations themselves. Renowned polar explorer Roald Amundsen—the first person to reach the South Pole and one of the first at the North Pole—suffered an accident as he attempted to assist Umberto Nobile. Nobile and Amundsen had flown over the North Pole two years before. In 1928, Nobile attempted the feat again with the airship *Italia* but crashed over the Arctic. Nobile and a few surviving crew members were rescued later on, but Amundsen disappeared without a trace; the wreckage of his aircraft has never been found.

Even now that technology has improved, and polar explorers are rarely lost, some expeditions require outside help to escape the clutches of the ice. In 2015, the twenty or so inhabitants of a Russian ice camp were rescued after just four months on their ice floe. After this, Russia abandoned the program of small drift stations it had launched in 1930; the ice has simply become too thin as a result of climate change.

We fly about sixty miles, with a full moon and an unbelievably clear view. The frozen landscape stretches out beneath us in the moonlight, with no end in sight. The ground view is limited, but up here we can see it all, suddenly confronted with the overwhelming scale of this frozen world. And then, all of a sudden, a light flashes, a tiny sign of life in this all-encompassing wilderness—the *Dranitsyn*.

December 12, 2019: Day 84

THE *DRANITSYN* HAS made little progress overnight. Will she actually arrive today? We arrange another day of research, recording measurements and operating the cities as usual. It'll probably be the last time we do this in the first expedition phase.

December 13, 2019: Day 85

THE *DRANITSYN* ARRIVED during the night! At 8:00 AM she begins the complex maneuver of steering her bow to our stern. Two hours later, the ship is next to us, still. What a remarkable meeting in the polar night—two of the world's best icebreakers, in the middle of nowhere, working toward another expedition milestone. The bunker hose is passed over to fill the *Polarstern* with fuel. Cargo operations also begin; there's a lot of freight to exchange between the ships. The *Polarstern* doesn't just need fresh food; the *Dranitsyn* has also brought additional research equipment for the coming months and will remove anything we no longer need to create space for the new cargo. It'll be several days before we're done.

December 18, 2019: Day 90

BY MIDDAY, ALL the freight has been transferred. We teach our new colleagues the complex procedure for lowering the large CTD rosette into the water via our onboard crane. The CTD rosette is a permanent fixture on the *Polarstern* but has never been used in these conditions. We have therefore devised procedures to keep sensitive instruments warm when we lift them out of the water and into the icy Arctic air. Plus, we use a totally different method to lower the CTD rosette through the ice hole and into the ocean: we hang it on one of the ship's cranes, rather than the slide beam, which normally lowers it over the side of the ship. This exercise is the last stage of the handover.

It's almost time to say goodbye. It feels strange. I will be relinquishing my onboard responsibilities for a time, organizing the rest of the expedition from the mainland until the middle of March. I still have no idea that a pandemic is coming, that it will turn our plans upside down and bring the expedition to the brink of failure.

Farewells are part of expedition life. New teams are constantly forming, friendships develop quickly and intensely, and before you know it, you're saying goodbye again. The more experienced participants are used to it. But this goodbye is different. Our team may be traveling home together, but we take our leave from the new arrivals with a series of hugs, surrounded by the endless black and icy wasteland close to the North Pole. We're also leaving our home of the last few months, the *Polarstern*, and the research camp into which we have poured our hearts and souls, that we have kept alive in the face of storms, cracks, and towering pressure ridges, that we have continued to expand, often working through the night. We bid farewell to our ice floe, where we have lived through so much and which we have come to regard as home—admittedly, a highly unusual home. What will it look like when I return in the spring and see the camp in the light for the first time, when polar day has commenced and the sun is a constant presence in the sky?

Emotional, we gather on the bow of the *Dranitsyn*, the area closest to the *Polarstern*. Music plays from the *Polarstern*'s loudspeakers and both ships toot their horns; then the *Dranitsyn* shudders and we move slowly backward. The decks of both ships are a sea of waving arms and shouted final messages. Gradually, the *Polarstern* grows smaller, and we can't hear her anymore.

The captain skillfully navigates the *Dranitsyn* backward along her approach route, aiming to disrupt the ice around the camp as little as possible. After a while he turns the ship, creating the space she needs in the ice. We are on our way.

December 19, 2019: Day 91

THE FIRST DAY of the return journey begins with hard ice, permeated by thick pressure ridges. We're forced to ram the ice over and over again. We may be heading back to civilization, but it's going to take weeks. In winter, the Arctic is one of the world's least accessible regions. It takes longer to reach the inhabited world from here than almost anywhere else on Earth. Thick ice barriers block the way, the distance is too great for helicopters, and airplanes can't land on the sea ice in the polar night. It'll be three weeks before we set foot on land. Even astronauts on the International Space Station get home quicker than this; their emergency escape pods can reach Earth in a matter of hours.

At first, the lights of the *Polarstern* peek through gaps in the fog. Then it goes dark.

December 20, 2019: Day 92

WE FIGHT THROUGH the darkness, alone. At 6:00 AM we get stuck. There's thick ice beneath the front of the ship and it refuses to break; we can't even dislodge the ship by directing our power backward. A huge ice ridge blocks our forward route. The instrument on the bow tells us it's twenty-eight feet thick. Eventually, even this massive barrier gives in to the mighty *Dranitsyn* and we travel a few more yards toward home—until we meet the next massive ridge.

This continues for the next few days; our progress is slow. We get stuck in the ice many times and in some cases wait for days until the ice pressure around us abates and we can free ourselves. This is normal when traveling through thick sea ice.

When the ice pressure is too great, there's no getting through. A smart captain will wait it out rather than wasting fuel—a precious resource on such a long journey—on a futile struggle against the natural power of ice compression. I have often spent days simply waiting in the ice. I've warned the expedition members who are less experienced in ice travel that this cycle of waiting and slowly moving is all part of the process. It's easy to feel nervous when you're trapped on a ship in the vast Arctic, too far away for outside assistance, making no visible progress.

Today we received a distress call from the *Lance*, a Norwegian research ship. She's stuck in the ice north of Spitsbergen and can't get free; the crew want to know if we can help them. The *Lance*, for her part, had set out to rescue two adventurers who decided to cross the Arctic ice on skis in winter, got into trouble, and realized they didn't have enough supplies to reach the ice edge. The *Lance* rushed toward them into the ice, sent out a rescue team, and managed to find the pair the day before their provisions ran out. They're now on board the ship, all stuck fast together, although they do have plenty of supplies and a safe, warm, and stable ship.

In the polar regions, you never hesitate to help others. Polar travelers have always been this way. We would sail to the *Lance* immediately, even if it meant delaying our return by weeks, but our Russian captain calculates that we don't have enough fuel to get there. At best, what we have will get us out of the ice, and only if we choose the direct route west of Franz Josef Land; even then, we'll be cutting it close. I speak with the *Lance* several times via satellite phone and explain that the *Dranitsyn* won't be able to help until we've reached the next harbor and taken on more fuel. And that will take a while, several weeks in fact. Given the ice conditions now that winter's setting in, we can't say how long it will be. And no other icebreakers are available; the *Lance* will have to sit tight in her dark prison. In January, the *Lance* manages to free herself from the ice—before we've even reached land ourselves— after the wind improves and reduces the surrounding ice pressure.

Slowly, we settle into life on board the *Dranitsyn*. After the constant, intensive work in our ice camp, we feel the tension drain away. Many of us spend most of the first two days asleep. Then we start to fill the sports room and sauna and get used to life on this new ship. We soon have a routine of Russian cabbage soup at midday and in the evening; otherwise, we could quickly lose all sense of time, in the darkness with nothing to do.

And Christmas is just around the corner! Secret Santa presents pile up in my cabin. Back in September, weeks before the expedition began, I asked everyone to bring little Christmas gifts. We're getting creative with the gift wrap, improvising with anything we find on board—old maps, aluminum foil, toilet paper. Some people even brought wrapping paper from home, just for the occasion.

December 24, 2019: Day 96

IT'S CHRISTMAS EVE. For Germans, this is the most important day of the festive season. Christmas can be a difficult time on an expedition; we miss our families even more than usual, and melancholy thoughts can easily spread, so a sense of community is more important than ever. And how better to fight the sadness than by celebrating? The ship doesn't have the supplies for a traditional Christmas party, so our team does what MOSAiC does best—improvise!

Russians celebrate Christmas on January 7, so December 24 means nothing to the crew. There's no special food today, but the *Dranitsyn* has brought stollen, chocolate Santas, and traditional festive cookies from our fantastic logistics team. They've even sent three Christmas trees. We spend the day decorating the ship and the Christmas trees in the Russian crew's mess, our mess, and the bar room, where we will later hold our Christmas party. We adorn the trees with fairy lights and quickly crafted Christmas decorations; great fun is had all around.

Christmas Eve kicks off in the afternoon with a festive Advent coffee and plenty of stollen and cookies to get us in the mood. In the background, the singers and musicians among us practice their Christmas songs for the evening.

After coffee, we break for our evening meal, and then the Christmas party begins. To our great delight, the Russian kitchen team have prepared mulled wine for us, even though they aren't celebrating. Once everyone is fed and watered, I give a speech, just like every expedition leader before me. I may be joking slightly when I reveal our true, secret objective:

> It's Christmas! ... I'm sure you're expecting me to talk about our successes, about the amazing research camp we built for the next team.
>
> But is that true? Is that the whole story? To be honest with you, no, it isn't! The expedition was a disaster, a total failure. We haven't made any progress with our covert objective!
>
> Did you really think MOSAiC was about climate research? That's just what we tell the public. In fact, we are on a secret mission, a mission many of you don't know about. Our mission is to solve no less than humanity's greatest mystery: Does Santa Claus really exist?

Since our efforts to date have clearly been unsuccessful, and there have been no scientifically proven Santa sightings, we speculate that he must be busy delivering presents to households around the world, including our families. Of course he wouldn't be in the North Pole tonight, putting his feet up! But that means our project is not yet complete:

> Today we celebrate Christmas Eve in Santa's homeland. Many of us will be returning here in spring or summer. I'm sure we'll see him then, lying in a hammock at the North Pole, relaxing on an ice floe in the constant sunlight of the polar day, with his white beard, dark sunglasses, and a cool drink in his hand. And then MOSAiC will be a success!

After my speech, we all sing Christmas songs and enjoy performances by our musical colleagues, some of whom have even brought along their instruments.

The music has barely finished when the door flies open to reveal Santa Claus himself, along with his little helper, Knecht Ruprecht! Their outfits look authentic, and nobody has seen them on board before. How can that be? Everyone is rendered speechless. Does Santa Claus really exist, and has he used his sled to visit us in the endless ice?

There's a burst of laughter, and soon the whole room is howling. The clean-shaven Knecht Ruprecht, never before seen on board, is actually our Finnish colleague Jari Haapala, who usually sports a luscious white beard that would put Santa himself to shame. This selfsame beard is now affixed to the previously hairless face of our young colleague Mike Angelopoulos, lending him a truly impressive Santaesque appearance. The two men are practically unrecognizable. Knecht Ruprecht, who will remain clean-shaven for the rest of the expedition, begins to tour the room, grimly asking everyone whether they've been good. Soon he is satisfied and opens a sack of presents to be shared among us all.

The mulled wine makes us feel cozy and warm inside; we sing some more, and our Christmas party continues late into the night.

We've organized a Christmas schedule of films, talks, and improvisational theater. The days fly by. We all miss our loved ones—I know I do—but there's plenty to distract us, and we enjoy a very pleasant Christmas overall.

December 26, 2019: Day 98

ON BOXING DAY, the black and cloudy sky opens up to reveal the stars, a sight that never fails to overwhelm. And at midday, the Arctic begins the spectacle it does so well. For the rest of the day, the magnificent northern lights paint the sky above us; as the sky grows seemingly larger, we feel increasingly insignificant. This is the first sign that we are returning to land: we have once again entered the latitudes at which the aurora is most frequent and intense. We were too far north before. And again, we can clearly see that the North Star—which always points north in the northern hemisphere—is no longer directly above us. We are getting closer to home.

But progress remains painfully slow. Again and again, the only way for the ship to forge a path is to ram the ice back and forth. Sometimes we cover just a few nautical miles per day. We continue south and enter the region north of the Arctic Cape, where ice tends to accumulate off Severnaya Zemlya. Once we have passed the cape, we will head west, and hopefully conditions will be better. But there's no way of knowing when that will be and when we will get home. We'll definitely be celebrating the New Year on board. My younger son turns ten on January 7 and wants nothing more than to see me again. After failing to be there for Christmas, I'd love to fulfill his wish. But that's up to the ice.

December 27, 2019: Day 99

THE ARCTIC CAPE is behind us! We are south of Franz Josef Land, heading west, still in thick ice but making better progress than in previous days. Yesterday I saw a bird in the ship's headlights for the first time in months, a herald of the ice edge, the thing we've been yearning to see for so long! It's an unmistakable sign that we are slowly leaving the grim hostility of the ice and moving toward friendlier climes. Daylight must be out there somewhere. Light that turns on and off without a switch, light that gives life and warmth, that separates day from night. And even farther south, the sun. We can hardly imagine seeing the sun, but we all spend our time picturing it, dreaming about it. We talk about it as something long forgotten, about times gone by when the sun still existed. And then it happens—at 9:00 PM, the ice no longer closes in around us. Our three searchlights reveal just a few ice floes floating in the open water, their patterns constantly changing in the lazy groundswell. Immediately the ship moves with the sea, and we return to the familiar rocking that was missing from the ice. I spend most of the night sitting by my cabin window, watching the floes and waves. Sometimes the ice is thicker, sometimes open water dominates. Birds soar above it all, enjoying the updraft created by the ship. And then, after three long months, we exit the Arctic ice.

December 28, 2019: Day 100

WE HAVEN'T RECKONED with the weather. We've barely left the ice when the wind freshens and the *Dranitsyn* begins to roll violently. "Rolling" is when the ship rocks gently along its longitudinal axis; for many people, it's more unpleasant than "pitching" (the harder up-and-down motion of the bow and stern).

The *Dranitsyn* is optimized for breaking ice, down to the very last screw. She's less comfortable in open water, throwing herself violently from side to side when the waves reach a mere ten feet.

And there's a storm brewing in the Barents Sea. We return to the ice for protection from the waves. We assume a waiting position and consult the meteorologists every day; we need a window of favorable wind to safely cross the open Barents Sea. We're cursed. First the ice blocked our path, now the open sea won't let us through.

Forecasts suggest there will be just one day with lower wind speeds and acceptable wave heights in one stretch of the Barents Sea. We want to seize this opportunity. On December 31, we will attempt to cross the Barents Sea and arrive in Tromsø on the evening of New Year's Day.

December 31, 2019: Day 103

WE SPEND THE LAST DAYS of the year planning for New Year's Eve, when we hold one last party on a rolling, wave-dancing ship. On the morning of January 1, the New Year greets us with our first daylight for quite some time; soon after, we see the Norwegian coastline. One hundred and four days after we left, we can see land once again!

Part III

ON LAND

▶ The Alfred Wegener Institute medical service tests expedition members for COVID-19.

7

ON A
KNIFE EDGE

Time to Take a Breath...

I'VE BEEN AT HOME in Potsdam for a few days now. It's strange—every time I return from an expedition, after months in a whole other world, gaining new experiences and making new friends, I find I've changed a little. They never fail to leave their mark. Meanwhile, it seems as if nothing has changed back home. Those few months have passed largely without incident. What felt to us like an entire stage of life was nothing more than a regular summer, fall, and winter.

Often, people returning from an expedition for the first time are surprised at how difficult it is. Coming back is much more difficult than leaving. There's an overwhelming joy at being surrounded by friends and family, and you can't wait to tell them all about it—but soon you realize that your experiences aren't easy to describe. No words can do justice to the things you saw, the sensations you felt. They just don't translate. Your thoughts keep drifting back to the ship, and the only people who know what you're going through are with their friends and family too, coming to the same, curious realization.

This has happened to me many times. I've dubbed it the "re-adjustment phase." It's different every time, and for every person. My expedition colleagues might be the only ones who can truly understand. It's peculiar; there's no other way to describe it. For months, you've only had contact with people you know well. And then you go to the supermarket, exchange a few words with a total stranger, and never see them again. It feels incredibly odd at first. On the expedition, you make do with what you have and improvise where necessary. The world is small, simple, and intense. Back home, the world expands; it's large, complicated, and yet banal. The normality of mainland life can be overwhelming. It takes time to settle in again.

Our little house in Babelsberg hasn't changed, exuding an almost impudent normality that I can't feel. I don't really try to tell my family about my experiences; I already know it would be tricky. Instead, we go for long walks and simply enjoy each other's company. I'll show them my pictures at some point and then transition into my stories. But that won't be for a while. At Easter, we'll have our traditional Christmas meal (goose) that I missed so much on the ship. It'll be a wonderful evening! But there's a lot to do before then. Just five days after returning to Germany, I don my expedition leader hat again. Life—and research—on the *Polarstern* continues. And the next challenges have already begun.

... And Face Something New

JANUARY SEES THE start of a huge logistical operation to transport the next team to the *Polarstern* and deliver provisions that will be essential in the Arctic winter.

The *Polarstern* has made good progress. She has drifted a long way and is less than 180 miles from the North Pole. On February 24, 2020, the drift will carry her to 88°36', farther north than any ship has ever reached in winter! Temperatures of −25°C (−13°F) are now normal on our floe, and they occasionally fall below −40°C (−40°F), which can feel like −60°C (−76°F) in the wind. Despite the extreme conditions, research continues steadily.

Shortly after returning, I start to see newspaper reports of a new virus spreading through China. It's somewhat alarming and yet far removed, like a distant rumble of thunder. Obviously, we're worried for the people of China, and feel for the scores of people who are falling sick and can't get the care they need. But, ultimately, we can put down the newspaper and get on with our lives. It doesn't affect us. After all, despite initial concerns, previous epidemics like SARS and MERS remained largely regional problems. And yet I can't help feeling uneasy. But I have no idea just how hard this thunder will hit us, that it will bring the entire MOSAiC expedition to the brink of disaster. And we have other problems too.

On January 28, the *Kapitan Dranitsyn* leaves Tromsø on schedule to exchange goods and people with the *Polarstern*. The operation runs into trouble straightaway. Since the start of the year, the northern hemisphere has been experiencing unusual weather conditions: a particularly strong westerly jet stream that is blowing around the Arctic. It's stronger than has ever been recorded at this time of year, and records date all the way back to 1950. Winds in the central Arctic are causing the ice to drift faster across the polar cap and directing all Atlantic storm systems straight to northern Europe. A storm is raging in the Barents Sea, and the churning waters are no place for a ship like the *Dranitsyn*. She's built to sail through thick Arctic ice, not through squalls. The crew have no choice but to take sanctuary in Tromsø's fjord and wait for better weather. On the plus side, it allows them to take on another shipment of toilet paper; there was a slipup when the cargo was first loaded onto the ship. Of course it had to be toilet paper, later to become a symbol of irrational worries (both in Germany and worldwide) as we realized the true scale of the coronavirus crisis. But we aren't hoarding unnecessary supplies; we are fulfilling the very real needs of a team isolated in the Arctic. It wouldn't ruin the expedition— there's always a solution—but they might as well rectify the error while they wait for the storm to pass.

After a nerve-shredding wait of almost a week, the water is calm enough for the *Dranitsyn* to brave the open Barents Sea and quickly seek shelter in the ice.

Our plans border on the impossible—to break through the central Arctic ice in the depths of winter and head for the North Pole region. By February, the ice is so thick and solid that it would challenge even the world's best icebreakers. Nobody has ever attempted to penetrate so far into the ice at this time of year on a self-powered ship.

The ice edge—where the open sea ends and the icy expanse of the Arctic Ocean begins—is much further south in the Barents Sea than is usual for February. The ship reaches the edge quickly, a consequence of the rapid ice drift of these last few weeks, which pushes the ice into the Barents Sea. It'll be slow going from now on as the ship breaks the ice. The *Dranitsyn* chooses the route between Spitsbergen and Franz Josef Land, meeting the thicker two-year-ice north of the Siberian island group. The only way for the ship to continue is to constantly ram the ice. Time and again, the ice pressure forces the *Dranitsyn* to stop, sometimes for hours, sometimes for a day or two. There's no point in fighting the ice pressure when it reaches this level. The *Dranitsyn*'s highly experienced captain does what any good icebreaker captain would do: he lets the engines idle and waits until the pressure abates. It takes decades of experience to develop these phenomenal skills. He reads the ice before him, identifies the weaker areas, and pushes forward, frees the ship over and over again, and continues his relentless march to the north. But weak spots in the ice are rare, and there are hardly any channels between the ice floes in winter. Some of the onboard team grow restless and worry whether they'll reach the *Polarstern* at all.

On land, we stare, spellbound, at the two little dots on our ice maps that mark the positions of the *Dranitsyn* and *Polarstern*. Progress is slow, but steady. Each day, the *Dranitsyn* closes the gap by around twelve miles on average. Occasionally her progress is halted for a day when the ice pressure becomes too great. We know that these enforced breaks are totally normal and there's no need for concern; there's no mistaking that the two ships are getting closer. We have no serious doubts about the success of the operation. Ice travel requires patience and calm. We calculate that the *Dranitsyn* will reach the *Polarstern* around the end of February, two weeks later than our extremely rough schedule; when it comes to ice, there's no point making specific plans. It takes as long as it takes. We lost around a week to the storm in the Barents Sea, and another to the tricky ice conditions; in early 2020, the ice in the Barents Sea is 50 to 100 percent thicker than in any winter of the last decade. This is due to the unusual winter weather and the accordingly rapid ice drift into the Barents Sea.

As the *Dranitsyn* closes in on the *Polarstern*, we realize that her fight against the ice has consumed so much fuel that she won't have enough left for the return journey. Her bunker capacity can't accommodate such an amazing feat. The *Dranitsyn* will be known as the ship that traveled closer to the North Pole than any other ship in winter, but what use is that if she can't get out of the ice again?

At this point, the *Polarstern* has enough fuel to last the next few months and doesn't need to take any from the *Dranitsyn*, but she doesn't have enough in reserve to fuel the *Dranitsyn*'s return. We need to find a solution before the *Dranitsyn* reaches the point of no return. Otherwise, our supply ship will be forced to abandon her task halfway through.

Our on-land logistics group are working their fingers to the bone. This is another perfect example of the importance of friends, particularly when you're operating in the polar regions, pushing the boundaries of possibility and constantly at the mercy of unpredictable weather and ice. Polar research has always forged close international relationships. The *Polarstern* has rushed to the aid of other ships so many times in the Arctic and Antarctic. Now, our Russian friends offer their immediate support. The *Admiral Makarov*, a huge icebreaker, is ready and waiting in Murmansk and can join our mission right away.

The two captains do the math and search for a place in the ice that will leave the *Dranitsyn* with enough fuel to get home and that is within range of the *Makarov*'s fuel reserves, even after transferring hundreds of tons of fuel to the *Dranitsyn*. They manage to identify an area within both ships' ranges. The amount of fuel to be transferred to the *Dranitsyn* must be precisely calculated so that both ships can escape the ice.

▶ HOW THE HOLE IN THE OZONE LAYER DISAPPEARED

The vast, uninhabited Antarctic was the last place you'd expect to find a hole in the ozone layer, but there it was. Unwittingly, my parents' generation had damaged the ozone layer with the halons in fire extinguishers and the hydrofluorocarbons (HFCs) long used in spray bottles, refrigerators, and plastic foam. The good news is that the hole will close in the second half of this century. This is all thanks to the Montreal Protocol (and the later tightening of its provisions), which compels practically every country on Earth to cease production of HFCs and halons, and was negotiated under the leadership of the United Nations.

This shows that even global environmental problems can be solved with timely and emphatic political action and a long-term, multigenerational mindset. Over a million annual cases of skin cancer have been prevented. Each generation needs to think about the ones to come, or such achievements are impossible.

We wouldn't have known the hole was forming were it not for the remote Antarctic observation stations. Without them, the news wouldn't be so good. Environmental early-warning systems cost a lot of money, so society needs to support them, to be in for the long haul. Otherwise, we won't be able to monitor our global environment as a whole or to identify and interpret warning signs in good time.

The *Makarov*'s tanks are filled and within days it sets off to meet the *Dranitsyn* in the middle of nowhere, surrounded by Arctic ice and the absolute darkness of polar night. We are now working with three of the world's most capable icebreakers: the *Polarstern*, from Germany, and the *Dranitsyn* and *Makarov*, both from Russia. There are now three dots dancing on our screens, an elaborate choreography that will allow the *Dranitsyn* to meet the *Polarstern* and then take on more fuel at the right time and in the right place. This operation alone will go down in polar research history. Nobody's ever done anything like this before!

As the *Dranitsyn* returns, its mission complete, another storm is raging in the Barents Sea. The ship waits in the ice for another week, finally sailing into Tromsø on March 31. It's such a relief to have completed this complex operation.

A New Virus—and the End of MOSAiC?

AT ANY OTHER TIME, sailing the *Dranitsyn* into Tromsø would be a matter of course. But everything has changed in the eight weeks since she left the harbor. Things are anything but normal right now. The ship and her passengers return to a world they could never have imagined.

That distant rumble of thunder didn't go away. It swelled into a storm and swept everything along with it. On January 23, China cordoned off the greater Wuhan area. Closing off a metropolitan region with over eleven million people? Unimaginable. Until now. It's like something from a cheesy disaster movie, a movie that will play out across the planet in just a few weeks.

Days later, the virus reaches Germany. The first cases in Bavaria are quickly identified and isolated. The population is lulled into a false sense of security, believing that our European systems will be able to contain every case straightaway. But many people are already asking what would happen if a recently infected person—still incubating the virus, but already contagious—decided to travel across Europe or attend a big party? Even the most forced optimism can't hide the trepidation that soon, we won't know what's hit us.

By the end of February, as both infections and deaths soar in Italy, we have a reasonable idea of what's coming. Some hospitals are hopelessly overstretched, with patient numbers far outstripping intensive care beds or ventilators. Doctors are forced to decide who will be ventilated and who will die, a living nightmare. The images of military trucks moving corpses out of Bergamo in northern Italy to relieve the burden on crematoriums will be seared on our collective memory forever.

And then the dam breaks, and I realize that even Germany can't control this, not even slightly. From now on, all we can do is react. News spreads that an infected person in Heinsberg, in western Germany, attended a carnival celebration. It's impossible to identify and isolate every contact of every person who was there. Despite the utmost—and highly commendable—efforts of the German health authorities, nobody can trace these kinds of infection chains. It's nobody's fault. Frankly, it was going to happen sooner or later. It's just that we all pushed it to the back of our minds.

Germany has one great advantage—a few weeks' warning—and can use the information coming out of northern Italy to make the right decisions, the same decisions being made throughout Europe.

On March 22, Germany introduces rules on contact with other people, "social distancing" becomes a buzzword, businesses close, and events are prohibited. Borders close all over Europe, including Norway, where the rules are particularly stringent—and where the *Dranitsyn* is supposed to drop anchor. What's going to happen to her passengers now?

As the crisis explodes around us and the entry/exit rules for various countries seem to change by the hour, our hundred-strong international team are on their way back to Tromsø. As the *Makarov* starts to transfer fuel to the *Dranitsyn*, Norway closes its borders. Ships can no longer enter Tromsø harbor as normal. One after another, European ports shut their doors. We quickly tell the *Dranitsyn* to take on more fuel in case we can't find a harbor to take her in and her odyssey is extended. The *Makarov* has sufficient reserves, but only just. Meanwhile, we frantically negotiate with the Norwegian authorities and come up with plans to get our expedition team onto land and back to their respective homes in the midst of the pandemic.

Negotiations are still underway as the *Dranitsyn* finally crosses the stormy Barents Sea and heads for Tromsø. A solution is approved by all sides just before she reaches the harbor: The *Dranitsyn* is permitted to dock and all expedition members (apart from the Norwegians) will be secluded from the outside world. They will immediately board a bus that will take them to an airplane we've chartered to fly them back to Germany. The Norwegians will stay in Norway and enter a fourteen-day quarantine. Germany doesn't require us to quarantine; the expedition started long before the outbreak and the team has been totally isolated ever since, so they're pretty certain to be virus-free (even when the ship was refueled by the *Makarov*, there was practically no personal contact). International team members can travel home on the few remaining scheduled flights from Germany. Coronavirus tests taken upon arrival confirm that nobody is infected.

And so the second-phase team return safe and sound—but we don't have long until the next handover on the *Polarstern*.

At our ice floe, light is starting to dawn, creating a totally different world from the one I left months ago. At the end of February, the *Polarstern* crew see a light on the horizon; on March 1, they experience something approaching daylight, and on March 12, the sun peers over the horizon for the first time. On land, we receive fascinating pictures of the ever-changing light conditions; they take me back to all those times when I experienced the first Arctic sunrise after the long, long night.

Our research is returning important results, even if they aren't always positive. Our research balloons detect a hole in the ozone layer above the middle of the Arctic. Never before has the ozone layer in the northern hemisphere so closely resembled that of the Antarctic. In the last few weeks, at the altitude where most ozone gas is located and forms a protective layer around our planet—about eleven miles— 95 percent of it has been destroyed.

While things continue as normal on our ice floe, one of the few places in the world untouched by the coronavirus, the pandemic wreaks havoc on our complex logistics.

I'm supposed to head back to the *Polarstern* in mid-March with our research airplanes, an advance party before the upcoming changeover.

We've planned loads of flights to the *Polarstern*; the aircraft are supposed to record measurements in the wider area to supplement those taken on the ice floe. We're also scheduled to land on the ice near the *Polarstern*, allowing me to return to the expedition. To reach the ship in the far north, the aircraft have to take off from Spitsbergen and stop for fuel at Station Nord, a little research station in northern Greenland that has a landing strip and aviation fuel.

It's Friday, March 6. I've packed my luggage and my flight to Spitsbergen is booked for Monday. The phone rings. It's Uwe Nixdorf. I can tell straightaway that something's wrong. A while ago, we decided that everyone involved in these flights would be tested for the coronavirus beforehand. Several of them met in Bremerhaven yesterday for their tests.

Uwe tells me that one team member tested positive and has been in contact with almost the entire group. Everyone who met in Bremerhaven is ordered to enter an immediate two-week quarantine. We won't be flying anywhere for the moment, that much is clear. Uwe and I decide to push the flights back by two weeks. I change my flight to Spitsbergen. This is just the start; the storm is about to overwhelm us, destroying everything in its path.

In the coming days, Spitsbergen is cordoned off indefinitely by the Norwegian authorities. The island is COVID-free and needs to stay that way. Its community of 2,500 people wouldn't be able to cope with an outbreak. Shortly after, Norway closes its borders to all travelers from outside Scandinavia. This is no longer a matter of postponing our aircraft operations—they've been rendered impossible for the foreseeable future.

The bad news comes thick and fast. We had planned to rotate the third and fourth teams in early April via flights from Spitsbergen with a Russian Antonov An-74 aircraft—that's gone out the window! In June and July, we've planned two journeys on the *Oden*, a Swedish research icebreaker, to take goods and fuel to the *Polarstern* and swap the fourth and fifth teams. A message from our Swedish partners bursts that particular bubble; the *Oden* has been ordered back to her home port and can't be used by MOSAiC while the pandemic continues. The Chinese research icebreaker *Xue Long 2* was supposed to deliver the final supplies and fuel and rotate the fifth and sixth teams. Our Chinese partners, too, inform us that the *Xue Long 2* has returned to port and is no longer available in the current circumstances.

New border closures are announced almost hourly. Eventually the entire Schengen Area shuts its doors to international arrivals, and most borders within Europe are closed as well. In just a few days, our entire logistical concept falls apart. We have a ship frozen into the Arctic ice and no idea how to cater for the second half of the expedition.

We always knew that MOSAiC might fail. It can't be ruled out on an expedition this ambitious. Accepting the possibility of failure is part and parcel of great achievements—not just in polar research, but in many areas of life. It's not a nice thought, but if you want to dream big— if you want to implement a project like MOSAiC and obtain urgently needed scientific observations of the central Arctic—then I believe you have to come to terms with it.

But nobody could have imagined that the expedition would be brought to its knees by a little virus that isn't even on the ship. Thanks to this virus, we now find ourselves on a knife edge. We don't know whether we can continue. We were aware that the expedition might falter, but it's hard to take now that it's staring us in the face.

It's looking bleak. For weeks nothing happens, and the travel restrictions and border closures just keep on coming. Public life grinds to a virtual halt in many of MOSAiC's partner countries. Within a matter of weeks, the world in which we planned this international expedition has all but disappeared. It's exasperating.

But we didn't come all this way to give up at the first sign of trouble. The polar region presents us with unexpected problems and challenges every day, and I guess we've been trained to be pragmatic and look for solutions rather than becoming mired in doubt.

Giving up is not an option. After each new blow, we pick ourselves up and try again.

First, we need to deal with our gigantic management problem. We may not know how to salvage the expedition, but we need the right structures to deal with the situation. We establish a streamlined, yet effective crisis response team comprising myself, Uwe Nixdorf (our head of logistics), and the two directors of the AWI, Antje Boetius (scientific director) and Karsten Wurr (administrative director). We need a broad base of support for the decisions to come. And right now, we need to expand the team that's trying to sort out this mess. These structures need to be up and running in a matter of days. And they are.

Our mission is clear. The logistics concept we developed so carefully for so many years is toast. We need a new one. And we only have a few weeks. The world has become a much more complicated place for this kind of international endeavor. Planning anything is hard right now because nobody knows what the world will be like next week, let alone in a few months. But it doesn't matter; we need to get to work!

All available logistical staff are mobilized. Despite her unbelievable workload, Antje Boetius rolls up her sleeves (her energy is truly astounding). Klaus Dethloff, who originally came up with the MOSAiC idea and drove the project for many years, cancels his imminent retirement.

The most pressing issue is our team in the ice. They have lots of fuel and provisions; the next supply run wasn't scheduled until June, on the *Oden*. But how much can you ask of a team isolated in the Arctic while the world falls apart? Many of them are concerned about their families and loved ones and would like to get home sooner rather than later.

What are our options? We consider abandoning the expedition so that the team can come home. But the ice is at its thickest and most solid at this time of year. The drift is moving fast, and the *Polarstern* has traveled farther than we would have expected toward the Atlantic and toward the ice edge there. Nevertheless, we're not sure the *Polarstern* has enough fuel to reach the ice edge without external help (she doesn't, as will later become apparent). The team will have to sit tight while we find a solution. The expedition has always been challenging to plan, and we've had to overcome many difficulties along the way, but this is the first time I've had trouble sleeping. My nights are restless and anxious.

While we work on a master plan to save the expedition, we also develop a small, targeted air operation to transport a few expedition members who cannot stay in the ice for various reasons. After several calls to many different places and authorities, Station Nord allows us to land and refuel once with two small Twin Otter aircraft, despite the risk to the station crew; if someone on board is infected, they could unwittingly pass on the virus when we refuel. Station Nord would never permit a larger operation—a big group of people would be far too dangerous—but we manage to complete our complex route with fuel stops on little airstrips in northern Canada and at Station Nord. On April 22, the Twin Otters (which always fly in pairs in case of emergency) glide down onto the icy landing strip prepared next to the *Polarstern*. They pick up seven expedition members and take the same route back to North America; from here, the returnees can travel to their respective homes. Our logistics group works tirelessly with the authorities to obtain the necessary exemptions from international travel restrictions. The whole operation is a masterful display by our logistics teams.

This buys us some time to find a way of swapping the rest of the team. The people on the ship say they are prepared to stay in the ice for several extra weeks. They continue their research, indefatigable. Now that it's spring, the temperatures are rising, and by late May or early June the *Polarstern* should be able to free herself from the ice and bring the team back.

The large safety buffers we incorporated into the original plans are paying off. The ship always carries enough fuel and food to survive the whole winter without further deliveries and wait until she's able to return by herself. We have made the gaps between supply cycles shorter than would have been necessary had the expedition gone according to plan. But when do things go according to plan in an operation like this? Unforeseeable events are always going to happen, which is why we always prioritize the safety of the team and ship when making our plans.

So the current situation doesn't constitute an emergency; the team are safe, warm, and well supplied as they wait in the ice to see what happens. Everyone on board knows that even if we don't find a solution, they'll be able to get back safely in the summer.

Now we need to focus on saving the expedition. If we can't find a way to make the upcoming deliveries, we'll have to abort by early summer at the latest—an unimaginable catastrophe for our scientific objectives. Winter and spring are done, but climate models need to cover the whole year. They need to accurately map the processes that take place in summer to create a more robust model of the Arctic climate system. If we stop now, there'll be a piece missing from the measurement time series the expedition is supposed to provide.

Over the next few weeks, most major expeditions and research ship deployments are aborted and canceled around the world. Meanwhile, we strive to save MOSAiC.

We sketch out every remotely feasible way to exchange team members in the depths of the Arctic and get enough supplies onto the *Polarstern* to keep her going until the end. We write down every idea, and soon we have a chart that looks like a huge tree. We contact all our partners and friends, in Germany and worldwide. Who has the resources, icebreakers, and ships (normal or ice-going), and who will let us use them, even in the current situation? Who will come to our aid?

Our Chinese partners get in touch at once; unfortunately, their ships are too far away. Our Russian colleagues are on the case straightaway; their two ships (the *Akademik Fedorov* and *Akademik Tryoshnikov*) are currently in the Antarctic but will be available in August, when the last team is scheduled to travel to the *Polarstern*. It looks like we've found a solution for the final supply run.

But as we tackle the upcoming changeover and supply run, every branch of our tree seems to be cut down. Most of our solutions turn out to be dead ends.

But not all of them! One possibility remains: two German research ships, the *Maria S. Merian* and *Sonne*, were forced to abort their expeditions due to the pandemic and are on their way back to Germany. The relevant authorities are quick to assist: the German Research Fleet Coordination Centre, which is managing the deployment of the two ships; the German Research Foundation, which operates the ships; and the German Federal Ministry of Education and Research, which owns them. The minister of education and research approves our solution right away. She has always supported our project, through good times and bad. The team members in the ice also agree, despite the great personal hardship it will entail: they'll have to remain in the ice for over two months longer and won't be able to support their families through the pandemic. This is the breakthrough we've been longing for! Finally, a plan is taking shape.

First, we reduce the number of planned personnel changes and supply runs from three to two: one at the end of May/start of June with the *Sonne* and *Merian*, one in August with the *Tryoshnikov*. We also need to adapt the scientific team—there will now be five expedition phases, not six—and clarify lots of other details.

The two German ships aren't icebreakers. The *Maria S. Merian* can operate around the ice edge, but definitely won't be able to reach the *Polarstern* in the depths of the ice cap. The *Sonne* isn't even remotely designed to work in ice; she usually operates in tropical and subtropical climes. But the exchange isn't scheduled until early summer; by then, the *Polarstern* will have drifted farther south and will probably be able to sail to the ice edge and meet the other ships. We plan for Germany's three largest research ships to meet at the ice edge and exchange team members and supplies. It's the first time these ships have worked toward a common goal: saving MOSAiC, the only major research project on the world's oceans that is continuing through the pandemic. It's a testament to the solidarity displayed by the research community and the authorities when the chips are down. And to our partners' efficient bureaucratic procedures; it normally takes at least two years of planning to deploy large research ships. It's all or nothing now, and the whole concept comes together in less than a month. We can truly be proud of the collaboration in the international polar research landscape.

The mission is saved! To say we're relieved is an understatement. After many weeks of worry, we have a solution and know that our work can continue.

There's just one last crucial point: we cannot allow the coronavirus to break out on the *Polarstern* or either of the supply ships after the next exchange. That would be a disaster; it would put everyone on board at huge risk and the expedition would fail. We work closely with the health authorities and come up with a plan. Everyone sailing to the *Polarstern* on the *Sonne* and *Merian* must undergo a strict quarantine of at least two weeks and pass multiple virus tests. The scientists and crew will spend two weeks in two hotels in Bremerhaven, totally cut off from the outside world. For the first week, each person will isolate in their room and have no contact with anyone else; in the second week, they will be allowed contact within their quarantine group, but distancing rules must be observed. After that, they will travel to the harbor, board the *Merian* and *Sonne*, and head north.

It's time for me to say goodbye as well. I will be returning to the *Polarstern* to lead the rest of the expedition. There's hardly any time to prepare, barely two weeks between the decision to use the *Sonne* and *Merian* and the start of the voyage itself. On May 1, I arrive at the hotel in Bremerhaven with my luggage. I'm about to surrender myself to weeks of confinement—and I'm looking forward to it, because it means I'm going back to the ice!

Part IV
SPRING

▶ Fourteen days alone in quarantine. A strange start to the next expedition phase.

8

RETURN
TO THE ICE

May 1, 2020: Day 225

THE DOOR CLOSES behind me and I am alone. I've never started an expedition quite like this before: over two weeks of quarantine in a Bremerhaven hotel, with each team member in a separate room. I check out the space that is about to become my whole world. This hotel was a good choice. On one side of the room, a floor-to-ceiling window facade reveals a small dock with research ships from other institutes. The open design makes me feel a little less confined and less cut off from the outside world. I'm sure it will help those who may tend to feel claustrophobic.

We are in perfect isolation. We don't see anyone, and we never leave our personal 250 square feet. Our doors stay shut apart from our three daily mealtimes. Someone knocks, we wait a few minutes, then we open the door, grab the food, and shut ourselves in again.

The internet isn't working yet. I lie on the bed. I have nothing to do. There's absolutely nothing to coordinate. All we can do is wait and see if anyone is carrying enough of the virus to be detected—then we can identify them and treat them. Right now, we are nothing more than human bioreactors, focused on giving potential infections enough time to erupt.

It's been years since I had so much leisure time. What a feeling! I enjoy it; I prepare a coffee, sit in the comfy chair by the window, and use my binoculars to take a virtual stroll around the dock. I've always enjoyed my own company and have no concerns about coping with quarantine. Quite the opposite, in fact. I take the opportunity to recover from the difficulties and stress of the last few weeks, from the near failure of the expedition.

I'll be gone for a long time now, around half a year. Yesterday morning I said goodbye to my sons, then my wife drove me from Potsdam to Bremerhaven in a sort of COVID-free private taxi. My elder son slept longer than usual and was sad to have missed part of his final morning with his dad, so we pushed back my departure and gave ourselves plenty of time together at home first—it's not like I have anything to do in Bremerhaven. After I said goodbye to my wife, she drove back too. Polar researchers' families have to be able to cope with farewells.

May 4, 2020: Day 228

THE INTERNET ISSUE has been solved. Our logistics department got hold of a few Wi-Fi routers and distributed them around the hotel; they're connected to the cell phone network, and now my phone and computer are flooded with emails. The coming days will be relatively normal for me, just like in the office. We also set up a WhatsApp group so that we can get to know one another and start to feel like a team. It works well; one team member will be spending her birthday in quarantine, so we arrange to serenade her from our windows. We also celebrate the results of our first coronavirus test. We're all negative!

The days pass faster than I expected. I haven't even switched on the TV yet. A lack of obligations is a luxury I don't enjoy very often. And the phone calls and emails keep on coming.

May 5, 2020: Day 229

TIME FOR OUR second coronavirus test. There's a knock at the door: Eberhard Kohlberg and Tim Heitland, the two doctors from our logistics departments, have come to swab noses and throats.

They aren't part of the quarantine group and could introduce an infection, so they wear full-body suits. We can only see a little bit of their faces. They are both experienced expedition doctors and have spent winters at our Antarctic station, which is how I know them so well. They designed this complex quarantine process, with staggered coronavirus tests for more than 150 people—the scientific expedition members and the crews of the *Polarstern*, *Sonne*, and *Merian*. Eberhard and Tim worked with the health authorities to iron out the creases and put the whole thing into practice in multiple hotels rented specially for the occasion. They have also arranged all exceptional permissions required for our international team members to enter the EU. It's a mind-blowing challenge and a logistical nightmare. It's thanks to them that we can continue the expedition safely.

May 7, 2020: Day 231

THE SECOND TEST is in, and we're all negative! This time we rejoice in the corridors. Given that we've all passed the second test, it's extremely unlikely that anyone in our group is infected. So now we begin the second quarantine phase: we are allowed to leave our rooms and mingle within our hotel, which is completely isolated from the outside world. We still maintain a distance of 6.5 feet, but we eat together in the dining room.

May 15, 2020: Day 239

THIS MORNING WE were swabbed for a third time, and around 11:00 PM I receive a call from Eberhard: all negative again! It's such a relief—we'd hug through the telephone if we could! Our efforts haven't been in vain, and we can set sail on Monday. I share the news via WhatsApp and the celebrations recommence.

May 18, 2020: Day 242

TIME TO GO! We travel by bus to the *Sonne* and *Merian*, which are waiting in the harbor to take us to the ice edge. As soon as we're on board—I'm

▶ Returning to the ice edge on two German research ships—the *Sonne* and the *Maria S. Merian*—to meet the *Polarstern*.

▶ A relaxing journey across the North Atlantic on the *Maria S. Merian*, heading to the ice edge in perfect weather.

on the *Merian*—we abandon all distancing rules. Now that we are proven to be COVID-free, we fall into each other's arms, an unusual sight after months of social distancing. Finally, we can greet our friends properly. Our team has bonded during quarantine.

Antje Boetius, Uwe Nixdorf, and Marcel Nicolaus have come to bid us a distant goodbye, along with a few family members. We wave to the shore and cast off at 11:00 AM. We spend an hour passing through the harbor area, then we're pitching across the ocean. We schlep our luggage to our cabins. On the morning of May 23, we aim to meet the *Polarstern* in the Is Fjord off Spitsbergen, an area near the ice edge that's sheltered from the wind and weather—perfect for the complex exchange between the ships.

May 20, 2020: Day 244

WE'VE REACHED THE NORTH ATLANTIC, but it doesn't feel like it. The sun shines on our faces, the sea is calm, and there are hardly any waves.

Many of us sit on the deck, knitting, reading, and chatting. Meanwhile, the *Polarstern* is fighting ice pressure and thick ridges pushed together by the wind. All leads have closed. Since she's making such slow progress, we reduce our speed to eight knots. We don't expect to see her in the Is Fjord before May 24.

May 21, 2020: Day 245

IT'S FATHER'S DAY in Germany. The weather's still fantastic and we're making good progress. We cross the polar circle at midday. Last night the sun set for the last time. From now on it will be a constant presence, our loyal companion as we travel farther north.

May 24, 2020: Day 248

THE SHIP'S MOVEMENTS became more pronounced yesterday, and the wind has freshened overnight. The ocean is churning around us, an occasional white crest on the waves as they wash over the working deck. The *Merian* is steadfast and remains pretty stable, but some people are seasick and the mess is much emptier at mealtimes.

Yesterday afternoon a group of whales joined us on our journey north, heading for their summer feeding grounds. For more than half an hour, we marveled at the clouds erupting from their blowholes and occasional glimpses of their backs and dorsal fins through our binoculars. Then they slowly disappeared behind the ship. At eight knots, we're moving a little faster than they are.

The *Polarstern* still hasn't made much progress and is fighting with the ice. Her new estimated time of arrival looks to be May 31. We'll have a long wait in the fjord, but we don't mind. This team understands that polar expeditions always depend on the weather and ice and can't really be planned.

May 28, 2020: Day 252

I CUT MY own hair today. There's no hairdresser on board, so I get busy with the scissors and mirror.

We've been waiting in the Is Fjord for three days. Even now, at the end of May, the *Polarstern* is hampered by the thick ice. If we'd been meeting earlier in the year, when she drifted much farther from the ice edge and the ice was much harder (on account of the season), she probably wouldn't have had enough fuel.

Since we'll obviously be waiting on the *Polarstern* for a while—at least a week—we need to make sure we don't succumb to cabin fever. And we need to do something about the space in the accommodation containers. The *Sonne* and *Merian* don't have enough cabins for almost a hundred scientists and crew, so we've set up containers on the decks of the *Merian*. It's four to a container; there are two little sleeping areas, each with a bunk bed, and a tiny bathroom in between. Inevitably, it starts to feel cramped after a while.

So I do two things: (1) I set tomorrow aside to convert the deck laboratory into a recreation room (a few people have suggested we do this). (2) We need entertainment, so I tell each team to find a creative way of presenting their scientific program—no slides allowed. One team will perform each night and the rest will make up the audience. The unusual format is born of necessity; we don't have a lecture room with a projector or screen because the *Merian* isn't designed for this many people. But it proves to be a stroke of luck and stokes our creativity.

May 29, 2020: Day 253

WE BAND TOGETHER and transform the deck laboratory into "Café Pallet" with two sofas made from pallets and yellow towels over the portholes for mood lighting. We get chains of lights from the crew, put comfy cushions on the sofas, and turn the lab tables into extra seating. A loudspeaker plays music from our cell phones, and there we have it—a super-cozy place to sit together, drink coffee, listen to music, and party in the evenings.

Café Pallet is extremely popular, and new elements are added all the time. The walls are adorned with watercolors of Spitsbergen's peaks, which we can see on the horizon. Plastic bottles are tied together to create decorations. We start using "pallet" as a byword for anything creative, improvised, and cozy.

June 1, 2020: Day 256

THE TEAM PRESENTATIONS are a resounding success. The Ocean team goes first. Since we don't have a meeting room, they transform the mess into the Arctic Ocean basin. In little groups, they simulate various bodies of water, acting out their movements and vertical layering in quite impressive fashion. They show models they've made and sing a song they've composed to explain the most important ocean measuring instruments. It's really fun.

Just when we think the Ocean team can't be topped, the Ice team present a three-act musical depicting the life cycle of an Arctic ice floe. They climb on top of each other to represent pressure ridges, the sun herself makes an appearance—sending her photons into and under the ice—and measuring devices (team members making beeping sounds) record the whole thing. The audience is moved when the ice floe dramatically fractures in spring, and the performance culminates in a brilliant, bellowing song (loosely inspired by "Bohemian Rhapsody") about an ice floe's inner life. Everyone's on their feet, and the standing ovations transition seamlessly into a birthday party for a member of the Ice team.

▶ BIG-BELLIED BOATS IN THE ICE

Many ships have reinforced hulls that enable them to keep moving through sea ice, but only a few dozen ships in the world can venture beyond the ice edge. Until the turn of the millennium, the number of icebreakers was dropping; the age of the great polar expeditions was over, and the Cold War was history. But now that global warming is making the Arctic more accessible, many countries are building new icebreakers, some for the first time. Russia has one of the largest icebreaker fleets, and is boosting its numbers with huge, nuclear-powered icebreakers such as the *Rossiya*, the largest that has ever existed. She entered construction in July 2019, is set to measure about 650 by 165 feet, and should launch in 2027. She will create channels in the ice big enough for large cargo ships, for example from the natural gas production sites in Siberia. The United States currently has just two icebreakers, but is planning more, and China recently complemented the *Xue Long* ("Snow Dragon") with the *Xue Long 2*.

After the pandemic forces everyone to rethink their plans, all assistance for the *Polarstern*'s yearlong drift is provided by Germany (the *Maria S. Merian* and *Sonne*, not icebreakers) and Russia (the *Akademik Fedorov*, *Kapitan Dranitsyn*, *Admiral Makarov*, and *Akademik Tryoshnikov*, its seasoned icebreakers).

June 4, 2020: Day 259

THE *POLARSTERN* APPEARED on the horizon this evening! She's moving closer all the time. We've been enjoying Café Pallet and the team performances so much that some people say (in jest, of course) that she could have stayed in the ice a little longer. We really have had a great time, and we hardly noticed the wait. There's no danger of cabin fever with this team around! The fantastic weather has allowed us to see all of Spitsbergen's mountains and glaciers. We've been constantly surrounded by whales—fin whales, humpbacks, and minkes—and all kinds of birds. The Atlantic waters are clearly filled with nutrients and food; where you find them, you'll also find the wildlife.

In the morning, we are joined in the fjord by the *Polarstern* and *Sonne*. It's an amazing sight: three huge German research ships in formation, the *Polarstern* in front and the *Sonne* and *Maria S. Merian* offset to the back, working together to drive MOSAiC forward, even in a pandemic. With the *Polarstern* so close, what long seemed an impossible dream is about to become a reality. Soon we will be on board. As major research missions are aborted all around the world, the MOSAiC expedition is able to forge ahead. There were many doubts, but now we can forget the pandemic and immerse ourselves in the Arctic!

Already waiting for us in the fjord is a tanker, which we'd arranged to supply fuel to the Polarstern. The bunkering begins, and the *Polarstern* takes on 2,800 tons of fuel. Our boats zip across the fjord, ferrying people between ships. The tanker, the *Sonne,* and the *Merian* take turns approaching the *Polarstern* in an intricate choreography. We also have cargo to swap between the ships. Soon we have everything we need for the tasks that await.

June 8, 2020: Day 263

IN THE MORNING we hold another video press conference with Anja Karliczek, the German minister of education and research, who wishes us all the best for our mission and the next expedition phase. She has stayed in contact the whole time and sends video messages at Christmas and Easter. Without her incredible help and support, we wouldn't have been able to continue through the pandemic.

We say our goodbyes and give little gifts to the amazing crew of the two ships that have been our home for the last few weeks, we drink sparkling wine, and then it's time to leave. We sail out of the fjord together, then the *Sonne* and *Merian* turn south and the *Polarstern* heads north. The ships toot their horns one last time, then our trusty partners disappear beyond the horizon, and we are alone again.

June 9, 2020: Day 264

WE'VE HAD A turbulent night; the *Polarstern* has been pitching in the strong wind from the north.

We reach the ice edge around midday. At first we find loose drift ice, a jumble of floes and water. We've all been waiting for this moment; we stand on deck and greet the white of the Arctic. We travel north-east along the ice edge for about two hours, then turn north, straight into the ice. The floes become thicker and form a closed surface that's still permeated by channels. And it's teeming with whales. One pokes his head out of the water right next to us. Perhaps he wants to check out this weird thing that's crossing his terrain, to see what it looks like above the water. Even whales are allowed to be curious. Then he sinks back down. He probably didn't like the noise. We cross the area quickly and leave him in peace.

June 12, 2020: Day 267

THERE'S NO WIND, and a thin layer of fog hovers above the ice with the blue sky shining through at the zenith. Even the sun gets through and bathes our icy surroundings in a pale, mystical light. We've had plenty of open water in the last few days, but that's over now. The farther north we get, the bigger the ice becomes. High pressure ridges shimmer through the fog in every conceivable shade between blue, turquoise, and white. Behind the white veil, piles of ice blocks resemble castles and mythical creatures.

The huge flock of kittiwakes that has been flying alongside us is thinning out. The birds realize that where we are going, they don't belong. Our destination is the central Arctic, with its vast fields of ice and snow. The birds live on the periphery of the sea ice, where they can hunt for fish in the fissures and channels.

The birds love the ship. As we break through the ice, we disturb the fish concealed beneath it and in the little hollows. Sometimes our huge

▶ Two kittiwakes fight over a fish roused from its hiding place beneath the ice by the *Polarstern*. They follow us in huge flocks, then turn back as we head deep into the ice.

hull flips entire ice floes upside down, little fish floundering on the top until a delighted kittiwake snaps them up a few seconds later. Dozens of birds have been playing this game for days, providing a soundtrack of relentless screeching. But now it's time for us to part ways. The birds can't survive where we're going, and our feathered friends turn back in their droves to friendlier climes. Gradually the landscape becomes silent and dead.

I sit on the deck and watch the ice structures pass by. After leaving the *Fram* and attempting to reach the North Pole, Fridtjof Nansen and Hjalmar Johansen skied across this area as they returned to the salvation of land. They had to drag their sleds over every ice ridge. The ice held promises of swift progress, but the landscape we've seen over the last few days must have been a nightmare. Nansen and Johansen could barely move for the endless labyrinth of channels, ice floes, and pressure ridges. They encountered multiple channels every day, and every time they had to lower their kayaks into the water, stow all their possessions, sail across, then lift their kayaks back onto their sleds—until they reached the next channel and did it all again. Eventually they gave up. They knew they couldn't make it. To save energy, they waited until late summer, their provisions growing scarce. Catching a seal bought them enough time until the ice was totally broken. Then they paddled to Franz Josef Land, their salvation. By contrast, we are pushing farther north! Our research camp is waiting for us on our ice floe, all alone, and we want to continue our measurements as soon as possible. We have made excellent progress during the night, so it's looking good. We've crossed the eighty-second parallel north and are now just forty-five nautical miles from the floe.

June 14, 2020: Day 269

YESTERDAY WE GOT STUCK (ice pressure) and now we're drifting. We tried to work against the ice for the first few hours, doggedly ramming the ship back and forth. Then we gave up and turned off the engines. We use the time to start some of our investigations here on the ice. I've been on the ice next to the ship with various teams. We've installed instruments and drilled ice cores. It felt so great to be on the ice again! We continue our work over the coming days and receive occasional visits from polar bears.

The weather is finally suitable for flying, and the MOSAiC floe is within range of our helicopter. I fly over to our good old ice floe with Matthew Shupe, who was also here for the first phase. I tell the pilot to fly in a wide arc over the floe and the thinner area where the *Polarstern* has spent many months since October 2019. This is now a huge shear zone, and the ice has broken into lots of little pieces. But the solid core, our old "fortress," is in excellent shape, stable with no cracks. The floe has delivered on its promise from last October—that the fortress would provide a reliable platform, even in summer 2020, to house parts of our research camp when the thinner surrounding ice began to fall apart. The sites crucial for studying the young, flat ice—where we have sampled and measured the ice continuously since the expedition began—are still connected to and accessible from the fortress. We will be able to continue our time series without any gaps! Even the devices that were left in the camp to collect crucial data by themselves are still intact on stable ice.

After gaining a general overview, I choose a landing spot in the fortress, and the helicopter descends onto the ice. Once Matt and I have disembarked, the pilot flies away, leaving us to explore.

▶ A visit from a polar bear and her cub, who is probably just a few months old. They both eye the strange thing that has appeared in their habitat but are otherwise unbothered by our presence.

What a feeling! We last saw this ice in the absolute darkness and ferocious cold of winter. Back then, anyone would have thought it came from another planet. In the meantime, this ice has drifted hundreds of miles through the Arctic and might still have some of our old, crusted footprints. But now we are bathed in the friendly sun of the polar day, the temperatures tepid at around 0°C (32°F). It's the first time we've seen the landscape in the light.

It's like an optical illusion. In my mind I see the winter night in three dimensions, but it's hard to marry what I saw then with what I'm seeing now. I recognize occasional pressure ridges that I used for orientation, but nothing really stands out—it's all too different now. I find a flag that I drilled into the ice in the black of night; it's still where I left it, and yet in another world. The feeling is indescribable. We're back in our camp!

I climb one of the high ridges and am rewarded with a stunning view. It's like hiking up a little mountain. From the summit, the frozen landscape extends to the horizon, white and gleaming. It's sublime to see it in the light and reminds you of your insignificance. The ice is endless and eternal.

When Nansen gave up on reaching the pole and first saw land—the mountains of Franz Josef Land—southeast of here on July 24, 1895, he reflected: "After nearly two years, we again see something rising above that never-ending white line on the horizon yonder—a white line which for millennium after millennium has stretched over this sea, and which for millenniums to come shall stretch in the same way. We are leaving it, and leaving no trace behind us, for the track of our little caravan across the endless plains has long ago disappeared. A new life is beginning for us; for the ice it is ever the same."[3]

How wrong he was! The ice may look endless and eternal, but it isn't. It has its limits and is growing smaller all the time. In the southeast, where Nansen wrote these lines, where he claimed that the ice would remain for all eternity, for millennia to come, there is now an open ocean. And it's only been a little over a hundred years. The ice has retreated. Human activity has warmed the planet to such an extent that it is no place for ice. How much longer will the ice survive where I'm standing now? Will my children's generation find the never-ending sea ice replaced by open water?

Then the helicopter returns, interrupting my thoughts. The pilots have noticed fog moving in; we need to leave immediately. We jump in and fly the forty-five nautical miles back to the *Polarstern*. We slalom through patches of fog as the light changes constantly on the ice below us, sunny spots alternating with white fog. Then the *Polarstern* materializes in the sweeping scenery.

The ship will reach our ice floe in three days.

9

THE
GREAT THAW

June 17, 2020: Day 272

AFTER A FEW DAYS of fighting with the ice and making no real progress, we finally see our ice floe before us. It stands out against the white and we can still see traces of our camp, instruments dotted around on the ice. At 8:15 AM, Captain Thomas Wunderlich—who has traveled with us and will stay until the end of the expedition—expertly guides the ship into the floe. Our colleagues celebrated their well-deserved return to Bremerhaven three days ago.

Our current, temporary mooring point allows us to explore the floe. What we find will help us figure out the ship's final position and direction. This approach worked well when we first anchored at the floe in early October 2019.

By 10:00 AM the gangway is lowered and we leave the ship. My team walks around the whole of the fortress to gain an overview of the ice structures and potential locations for rebuilding the camp; they packed up most of it before the *Polarstern* departed.

▶ The *Polarstern* arrives at the MOSAiC floe, this time in summer, in the constant light of polar day.

June 19, 2020: Day 274

FOLLOWING YESTERDAY'S EXPLORATIONS, we park the ship in her permanent position at around 9:30 AM. Last night we maneuvered the ship from the west side of the floe to the east and pushed a little farther into the ice. We still weren't satisfied with our position because the ice astern was all broken up and the ship wasn't stable.

This morning, we altered position slightly so that our starboard side is on the ice; the ship is currently wedged in by the ice floe and a larger area of ice on our portside. If the ice floe changes, we will use the bow and stern propellers to stabilize our position as we work on the ice during the day. Now that summer and the thaw are approaching, we can't use ice anchors to attach the ship to the floe—they would simply melt out of the ice.

Our floe now consists mainly of the old fortress, but there's also an L-shaped section of young ice on its western and southern side. This is the frozen lead that formed during the first phase and is still connected to the fortress. It's definitely less stable and might break eventually, but right now it offers excellent access to the typical one-year ice. It's also

▶ An ice structure on the edge of the floe.

one of our regular, key sampling sites. Since new ice first started to form in the lead in fall, we've taken continual measurements to document the entire life cycle of the ice from the moment of formation. These series of year-round measurements are what make MOSAiC so unique.

The old part of the floe is full of sediment and brown patches are visible on the surface, a phenomenon that will increase considerably as the summer thaw progresses. The sediment frozen into the ice collects on the surface as it melts, turning the floe darker and darker.

They may look dirty, but these brown spots are a natural process for the sediment-laden floes that come from the Laptev Sea and adjoining coastal seas. Storms stir up the sediment on the beds of the flat Siberian marginal seas, and it freezes into the sea ice as it forms. Essentially, the ice is a huge Arctic conveyor belt for sediment.

This conveyor belt is grinding to an increasing halt, with major repercussions for Arctic biogeochemistry. Our ice floe is now an oddity, a brownish island surrounded by lots of younger floes, which are white because their story did not begin in the Laptev Sea. Our measurements emphasize the contrast between the two, directly comparing the typical younger ice in our L-shaped area with the old core of the floe. The old and new worlds sit side by side—a stroke of luck for our research.

The dark surface at the core of the floe absorbs more radiation, so it's started to thaw earlier. This part of the floe stands out on the satellite images too; distinct melt ponds have formed already, sooner than in the neighboring floes and our young ice.

As the days go on, we find piles of stones all over the place, and even rocks over an inch in diameter. How do rocks like these end up on an ice floe in the middle of the Arctic, thousands of miles from land and 13,000 feet above the seabed? They prove that our ice floe originated in a flat coastal area where it had direct contact with a beach or the seabed, probably in the area around the New Siberian Islands. Some of the rocks are covered in large patches of algae, and mussels too. There's a striking contrast between the new one-year ice and the two-year ice in the fortress, and it allows us to examine the sediment's impact on the radiation above, in, and under the ice.

The atmosphere on board is excellent. We need a solid foundation for the difficult times that are sure to come. The ice might be nice and stable at the moment, but that is unlikely to last long.

June 24, 2020: Day 279

AFTER SEVERAL FOGGY DAYS, the clouds part and the blue sky bathes the ice floe in the blazing light of the low-lying sun. The summer solstice was four days ago, and the sun is high above the horizon (at least for the Arctic) where it circles around us twenty-four hours a day, burning down on the ice.

What an explosion of color compared to winter! After nothing but the black of the dark and the garish white of our headlamps, now the floe gleams in countless pastel shades. Melt ponds dominate the surface, from vibrant turquoise to a shimmering greenish blue and every color in between. There aren't enough words to describe them all. No

▶ CYCLING TO RUSSIA

When the researchers and crew returned to Bremerhaven on June 14, all sorts of travel restrictions had been introduced and most international air traffic had been grounded. Nevertheless, most of them made it home without major difficulties—apart from one Russian colleague from the AARI in Saint Petersburg. The next flight to Russia wasn't scheduled for another two weeks, so the AWI moved heaven and earth to reunite him with his wife and young daughter. An AWI staff member spoke to the border police, who assured her the man would be able to cross the Finnish/Russian border in a private vehicle. She quickly bought a folding bicycle on eBay and had it delivered to our Russian friend at Bremerhaven airport. He promptly flew to Helsinki—with his bicycle as special luggage—and then boarded a train to Imatra, a town on the border, arriving in the evening. As the daylight faded, he cycled around six miles to the border, on unfamiliar roads, with fifty pounds of luggage that kept falling off the tiny bike rack. He managed to cross over but was stopped on the other side by a border official. After five months of polar exploration, he possibly didn't look like a fine, upstanding citizen, and she wasn't entirely convinced by his plan to travel the rest of the way to Saint Petersburg in a taxi. He was left with no choice but to sit and wait. Four hours later, the woman finally showed mercy, called the long-awaited taxi, and our colleague fell into his family's arms at 3:30 AM.

two ponds are the same, and all the colors meld together in the larger ponds. Often we see pale ribbons where old cracks, long since frozen shut, run through the bottom of the ponds with areas of deeper blue and turquoise between them. I could watch the capering colors for hours. And above them all we have the yellow of the sun deep in the sky. The ice transitions from brilliant white to various tones of ocher and beige to light brown. The ochers complement the turquoise ponds beautifully, as though this colorful interplay has been carefully composed. If only we had a painter with us! And yes, the surface of the ice is increasingly brown with sediment. As I explore, I gather some of the stone piles for later analysis on land. When I spread out my treasure on a table on the ship, it looks like a pebble beach. Multi-colored pebbles, smoothed and rounded, with every conceivable shade and surface.

June 28, 2020: Day 283

THE LAST WEEK has been hectic. We've worked incredibly hard to rebuild our camp so that we can resume our research as soon as possible. Our decision last fall to use this special ice floe has been vindicated—our fortress continues to provide an excellent base for our camp, even in summer, while the ice around it has broken apart.

The ship is stable on the southeast side of the floe. The whole research camp is up and running and many of the old measuring fields have been activated. The camp is lighter and more mobile this time; we know that the ice is going to melt. Soon the surface will be a lakescape of melt ponds, and we need to remember this when we choose locations for our research infrastructure. Our new, mobile research cities allow us to respond to the dynamics of the crumbling ice floe as it reaches the end of its life cycle. All stations are connected to the power supply with mobile cables and to the ship via radio LAN.

We've arranged the camp along the spine of the fortress so that the equipment won't sink during the summer thaw. The main axis of the camp runs along one of the strongest pressure ridges, elevated and dry above the slowly emerging landscape of melt ponds. Here we have Met City with its resurrected meteorological mast and several outlying flux stations, which also measure turbulence in the lower air layer and are strategically placed on sleds for mobility. Then there's Ocean City, which has several points of access to the water column through holes in the thinner neighboring ice, and Balloon Town, where the hand-some white TROPOS tethered balloon has been floating since yesterday and Miss Piggy waits for temporary deployment. The remote sensing instruments have been installed at the Remote Sensing Site, also on mobile sleds. We also have smaller stations for seismic measurements and for extracting ice cores (in the same spot as before), snow and ice measuring fields, and much more.

ROV City has already sent "Beast" on a successful dive, and there's space to launch our drones over the one-year and two-year ice. The instruments on board, on the P-deck, in the crow's nest, and in the research containers at the bow are all running. We will probably begin full, regular operations tomorrow; the teams are scheduled to work six and a half days per week.

Both helicopters are ready to deploy, and the crew are excellent. The chief pilot, the onboard weather technicians, and the meteorol-ogist (we usually have a meteorologist on board, but she's assisting us from land at the moment) work so well together that we can run flights even in brief spells of good weather. Our chief pilot monitors the

▶ The *Polarstern* in the MOSAiC ice floe at the end of June 2020. Melt ponds have begun to form but have not yet melted all the way through.

weather constantly and gets straight on the phone to the meteorologist to make sure it's safe to fly.

Although our floe has already traveled a long way south—to 81°00′51″ north—it holds its ground. It's likely that we'll have to or want to head north at some point, but not right now. The satellite images show bodies of water between us and the North Pole. Later in the summer, we might find ourselves with an almost-open sea route to the North Pole—that would be unprecedented! If it comes to that, we should document it and seize the opportunity to study the processes that are, unfortunately, likely to shape the future of the Arctic.

July 2, 2020: Day 287

I'M TAKING A LONG EXCURSION across our ice floe. The great summer thaw is in full swing and makes its presence felt wherever I go. It's part of the Arctic life cycle. In winter, the ice covers large areas, then in summer it melts and retreats back into the central Arctic. This sea ice cycle is the Arctic's beating heart. Its rhythm has accompanied our planet for millions of years and can even be seen from the depths of space. Mars has a similar heartbeat: carbon dioxide ice forms at whichever pole is currently in winter (the pole turned away from the sun) and covers it in a white cap.

Wherever we go, the ice melts around us. Everything drips. Long icicles form on every ice ledge and in the shimmering, deep-blue ice caves in the pressure ridges. The usually silent Arctic now has a soundtrack, an incessant *plop, plop*. Some of the melt ponds have found their way to the edge of the ice floe, where meandering rivulets and burbling streams dig into the surface of the ice. Before long, the first ponds on our floe will melt through and create its first holes. Then the ponds will flow out beneath the ice.

▶ WEATHER REPORT FOR THE CENTRAL ARCTIC

Arctic expeditions are more reliant on accurate weather reports than almost anyone else. The weather can change on a dime, endangering teams on the ice and helicopters in flight. Unlike densely populated areas, this region isn't peppered with weather stations supplying data for regional forecasts, which is why the *Polarstern* has her own meteorologist and a weather technician on board. Several times a day, they receive satellite maps, results from weather forecasting models, and data from around the world (national weather services use all of these sources to compile their weather reports). We also have a team on deck who send up a weather balloon every six hours. The data from the balloon is fed into the global weather station network, and the *Polarstern* appears on the map as a lonely dot in the middle of the Arctic Ocean. The meteorologist compiles several reports a day using the data from the weather forecasting models and the weather technician's observations. Without these reports, nobody on the *Polarstern* could use the helicopters or travel far from the ship.

Meltwater from the ponds is now seeping through little cracks in the ice. A layer of fresh water is forming beneath the ice, distinct from the denser and colder salt water below it. The boundary between the two is astonishingly clear on the impressive images captured by our underwater robot—an even, lightly undulating reflective surface in the middle of the water.

And the fresh water is freezing under the ice! The underside of the ice is still −1.7°C (29°F)—the freezing point of salt water. But now there is fresh water there, which freezes at 0°C (32°F). It freezes into large needles on the underside of the ice. The ice holes we created in the winter for our instruments (and tried so hard to keep open) are freezing from below! The warmth released during the freezing process is an essential part of the energy budget. The freshwater layer was there before we drilled our holes, so we have to assume that it's created largely by natural drainage processes. However, we can't rule out the possibility that by creating our ice holes, we have accelerated these natural processes and contributed to the freshwater layer.

The air temperature is around 0°C (32°F) and the sun is shining. I wander through this pleasantly colorful and icy landscape and can hardly believe what I'm seeing. Half a year ago I stood on this exact spot and the darkness made me feel as though I were on an alien planet. Now it's warm and life is simple. No need to kit myself out like an astronaut just to go outside, no need to double and triple check that I've got my headlamp, a spare, and a spare for the spare (nobody wants to be plunged into the darkness of polar night). Now our trusty sun is in the sky, dependably lighting the way back to the ship around the clock. We end the day with a snowball fight on the ice.

July 3, 2020: Day 288

WE REMAIN STABLE in the ice floe. Polar bears visit occasionally, usually at night when nobody's on the ice. They all inspect the camp briefly before moving on; we don't need to chase them off to avoid habituation. Last night, at around 3:30 AM, a bear started biting the seat of a snowmobile; we scared him off with flares so that he didn't ingest the foam. It worked perfectly and he didn't try to come back.

July 5, 2020: Day 290

THERE'S LIFE ALL around us! There are birds in the sky all the time, and seals pop their round heads out of the open water. The macrofauna (i.e., polar bears) behave themselves on their (now constant) visits and move on quickly without causing major damage or eating any of the scientists. Giulia Castellani, one of our biologists, casts longlines into the ice hole, frequently producing large fish from depths of hundreds of feet. Initial analyses show that they are healthy and well fed. The fish kindergarten is right beneath our feet. Myriad tiny fish hide in the rugged cracks and caves on the underside of the ice. We see them on the underwater images and sometimes spot them in the cracks.

Who Lives at the North Pole?

POLAR BEARS CAPTIVATE anyone who has the privilege to see them in their natural habitat. I've been to various Arctic regions and have spent many hours watching them roam their territory or stare intently at an ice hole, waiting for hours in case a seal appears.

But polar bears are just the end of the long Arctic food chain that begins beneath our feet, in and under the ice. There are countless microorganisms swimming in the water—viruses, bacteria, and primeval archaea, which (like bacteria) have no nucleus and are found mainly in our planet's extreme environments. These single-cell organisms are all invisible to the naked eye, but we can see other species. On every dive, our underwater robot spots an ever-growing carpet of long green, brown, and orange algae on the underside of the ice, a two-foot-long beard wafting in the water. This consists mainly of *Melosira arctica*, extensive colonies of tiny diatoms that are also single-celled but merge to form chains. Together, they improve their chances of colonizing this habitat. Diatoms also produce substances that protect them against the salt in the alkaline channels formed by the ice as it freezes. One of the questions MOSAiC aims to answer is how well algae thrive in the low-nutrient water of the Arctic Ocean and whether they form similarly long carpets in the high north as they do farther south.

The cracks and fissures in the ice are teeming with other plankton too—tiny creatures (some of them microscopic) as well as larger copepods, shrimps, and little fish such as *Boreogadus saida*, or Arctic cod, which occurs farther north than any other fish. We watch the Arctic fish kindergarten on the robot's videos.

Phytoplankton convert solar energy into biomass. Zooplankton (such as crabs and shrimp) and young fish live on phytoplankton and also like to eat each other. There's an abundance of life beneath the ice, and now that it's summer, it's reaching its peak. And this is the basis for all the larger lifeforms around us. The fish eat the plankton, seals and birds eat the fish, and polar bears live on seals. Each building block has its place, and without it the system won't work. If one block is removed, the whole thing is jeopardized. The Arctic sea ice is as unique an eco-system as the tropical rainforest, just harder to access. And it threatens to vanish as our greenhouse gas emissions heat the planet and the ice continues to diminish.

▶ A polar bear inspects our camp and then continues on its way.

July 7, 2020: Day 292

I'M LISTENING TO MUSIC in my cabin—Pilleknäckeren—and writing my diary. Finally a quiet evening, the first in quite some time. The nighttime sun floods my cozy cabin as the ice outside lies silent and untouched; all our teams are back on board after a day of intensive work. Pilleknäckeren are a band from the Göttingen pub scene of the early 1990s. They split up a long time ago, but I took the only two cassettes they ever recorded—*Red* and *Yellow*—on my first voyage to the Antarctic in 1994. I listened to them on a continuous loop as I analyzed sam-ples in my container on the P-deck with icebergs and whales passing by outside. Since then, I associate this music with the polar regions more than any other. Like scents, music can evoke specific memories.

I'm still friends with some of the people I met on that trip. My best friend on board would have loved to be with us today, but her life took a different direction. While I'm away, she visits my family and yearns to be here in the Arctic. As I write these lines, I am reminded once again of just how privileged we are to explore this unique landscape, this extraordinary habitat. Our work is hard and the lengthy expedition is full of privation. But at no point would I swap places with anyone else in the world.

We are witnesses to a vanishing world. Will there be any ice in the Arctic as our children grow up? We can still do something about it. I can only hope that humanity comes to its senses in time and ends the madness of rampant greenhouse gas emissions. If it does, there's still a chance to preserve this icy world for our descendants. Otherwise, it will perish with our generation.

I banish my bleak thoughts and look at the fascinating landscape of ice and sun outside my window. It's still here, for now! And we have more immediate problems. How long will the ice floe hold? It's coming to the end of its life cycle. We are currently around seventy-five nautical miles from the ice edge. Now that it's summer and the ice is retreating, we are drifting toward it and it toward us. The floe is entering its final days. This summer, it will reach the ice edge near the Fram Strait, be shattered by the swell and waves, and finally melt.

When we enter the swell, it will be time to dismantle the tents on the disintegrating ice floe. After that, we have one more job to do. When we began our expedition in late September 2019, the freeze was already well underway. We missed the initial freezing phase, the point at which ice floes like ours are born. We want to rectify that now by studying the final phase of the yearlong Arctic cycle, the last piece of our puzzle. And where does the freeze begin, where can we focus on this process before completing our expedition in mid-October? In the far north! This will be our next destination.

Yesterday I talked to the captain about our fuel supplies. The floe's stability over the last few weeks has saved us lots of fuel. If things stay the same for the next month, we will have enough fuel to keep all options open for a journey to the far north. I gaze across the ice. It looks stable. Will it be a dependable home for the month to come?

Part v
SUMMER

▶ The soft light of a rainy day dulls the colors and blurs the contours of the ice.

10

HIGH SUMMER ON THE ICE

July 11, 2020: Day 296

IT'S RAINING. In summer, the Arctic's frozen winter landscape disappears, and temperatures increase to around freezing. This is totally normal and nothing to do with global warming. Ice forms and spreads in winter, melts and retreats in summer. It's always been this way: the eternal heartbeat of the Arctic, the rhythm of the seasons. And as long as the ice freezes sufficiently in winter, the central Arctic ice will outlast the summer thaw. But we are disturbing this sensitive cycle through global warming—in the future, the Arctic could lose all its ice in summer. If that were to happen, its heart would stop beating.

Right now, we have something in common with the people back home: hours upon hours of persistent rain. But today's a big day. All the teams are pulling together to complete one of our twenty-four-hour measurement cycles. From midday, we'll be recording data like there's no tomorrow! The rain isn't making it any easier. The snow turns to deep, heavy, water-soaked sludge; it's even harder than usual to move across the ice and we're gradually getting drenched. Our polar gear is optimized for low temperatures; slowly but surely, it soaks up the constant rain.

▶ Luisa von Albedyll on a nighttime polar bear shift during a rainy twenty-four-hour measurement cycle.

But our enthusiasm can't be dampened. At 12:00 PM on the dot, the teams descend on the ice and begin their precisely timed measurements under and in the ice, in the snow, and in the atmosphere. These twenty-four-hour cycles are a scientific treasure trove. No day in the central Arctic has ever been recorded in so much detail.

It stops raining in the afternoon. The sun even shows its face occasionally—and as soon as it does, the helicopter lifts off to take measurements from the air. A good omen for this crucial day! But just twenty minutes later, the pilot reports approaching fog and returns to the ship. Before long, the view from the bridge is almost entirely obscured. Visibility is around 200 yards, sometimes just 150 yards. We reconfigure the polar bear guards and adjust to the new conditions. If visibility drops any further, we won't be able to keep everyone safe from bears and will have to abort our work. But things remain stable and the cycle continues.

After completing my night shift on the bridge, I visit the teams on the ice at 5:00 AM. Visibility has improved, but it's still raining occasionally. The Arctic weather can change in an instant.

In Ocean City, under the watchful eye of the polar bear guards, a colleague lowers the turbulence probe (a little device with a mop of fuzzy orange hair) through the ice hole and into the ocean over and over again. It's a monotonous job. For safety and entertainment, they have a series of companions who keep the conversation going and make sure they don't nod off during the long night. Several people have volunteered for hourly shifts as an "ocean friend." It's kind of like icy speed dating.

Below the shapeless, uniformly light-gray sky, the light is extraordinary. The harsh contrasts of a sunny day have vanished, and everything is bathed in a wonderful, almost cottony soft light. There's a shimmering beneath the piled-up pressure ridges. Thick layers of frozen water take on an astonishing blue color, and a deep blue light emanates from the icicle-filled caves.

Luisa von Albedyll, a sea ice researcher, is guarding her Ocean team colleagues from polar bears. Everyone on MOSAiC takes their turn to help each other out, even in the middle of the night. I chat with Luisa for a while, taking in the beautiful ambience, before moving on to the Eco Lodge.

The ecosystem researchers regularly spend twenty-four-hour cycles in the Eco Lodge, sampling the life beneath the ice. They take samples at specific intervals, and the rest of the time they just try to stay awake. They've connected both of their tents to the power supply and set up a coffee machine to create a cozy café on the ice, with a little ice hole in the bottom to facilitate the science. There's a cup of coffee waiting for me when I arrive—my colleagues heard me radio the bridge to tell them where I was going. The bridge watch is always informed about everything, even nighttime coffee stops, so that nobody gets lost. Despite the long shifts, the mood in the Eco Lodge is excellent, and the smell of coffee is a real pick-me-up after a night without sleep.

But that's not all the Eco Lodge has to offer. When the windows are closed in the bottomless, lightproof dome tent, the light backscattered from the ocean permeates the six-foot-thick icy floor. The ground glows an intense deep blue and creates a magical atmosphere.

Work on the ice continues late into Sunday afternoon. In the evening we celebrate our successful cycle in the Zillertal.

July 12, 2020: Day 297

OUR CELEBRATIONS COME to an abrupt halt. Shortly before midnight, the ship shudders and the ice pushes us forward by about one ship's length. The power lines on the ice move and we send out a team to intercept them and detach them from the ship. Looking around, we can see that the ice floe itself is unharmed and none of our instruments are in immediate danger. We fire up the *Polarstern*'s two additional engines so that the ship has more power to maintain her position against the pressure from the stern.

And then the fire alarm sounds on the bridge! Throughout the ship, the automatic fire doors close as the ship protects herself and prepares for the worst. A fire on board is by far the greatest danger for this ship. It is our ultimate nightmare; without the ship, we would lose our secure base and be stranded in the Arctic with brittle ice and miles of water beneath our feet.

In the fire control center on the bridge, the smoke detectors show smoke developing in the rudder engine room and a neighboring room. This is no error, this is serious—two smoke detectors wouldn't return identical messages if there was nothing wrong. And at the same time the rudder refuses to move. What's going on?

Apparently, the ice pressure has pushed a massive block of ice under the stern and against the enormous rudder, overloading the rudder engine. The hydraulic pumps couldn't withstand the pressure, the gaskets have blown, and hydraulic oil has spurted onto the hot adjoining pipes. The oil has vaporized and created the "smoke" detected on the bridge. The engine watchperson dashes to the rudder engine room and radios a status update. After a brief discussion with the chief—the person in charge of the engines—the captain switches off both rudder engines for safety's sake. Without them, we are unable to maneuver the ship.

Here in the ice, we can take the precaution of switching off the second, intact rudder engine without causing problems. Ice can be extremely troublesome, but it can also protect a ship.

▶ The two tethered balloons float above the summery ice floe on July 13, 2020.

Throughout the history of polar exploration, countless ships have been squashed in the ice, claiming the lives of their crew. For modern icebreakers, the sea ice is a haven; their hulls can easily withstand the pressure. Many icebreakers are not well suited to sailing in open water, rolling and pitching violently in even the slightest of swells. The *Polarstern* is different, a world-class icebreaker who acquits herself exceptionally in open water; this is partly due to her two stabilizers, little adjustable wings below the water that dampen her rolling motion. But right now, we're glad to be protected by the ice. In the open water, a ship that can't be maneuvered or steered would be at the mercy of the waves, wind, and currents. In the ice, we can wait patiently for repairs.

By the morning, the damage to the rudder is rectified, and I discuss our next steps with the captain. We need to return to our old position, where the power lines end and the center of our camp is located. We can't go back because of the huge blocks of ice behind us. We could use the *Polarstern*'s enormous bow to push them out of the way, but this will only work if we first proceed forward, make a large loop, and then steer the bow around to return to our previous position. The problem is that we've set up loads of stations on surrounding ice floes on our portside, and they mustn't be damaged. We don't even know where some of them are because their GPS buoys stopped transmitting ages ago.

I use the brief gaps in the fog to locate the old stations from the bridge and note down their direction and approximate distance. It's too foggy to measure precise distances with the laser devices. But I can use the data to identify the floes on which the devices are situated and mark them on the ice radar. Now all we need to do is plot a course through this labyrinth of ice floes and old measuring stations to return to our previous position without damaging any equipment. Before long, the ship is moving with the full force of her four engines.

While I monitor the ice radar to make sure we don't hit our measuring stations, the captain maneuvers the ship through the ice, skillfully negotiating the tight curves without damaging a single station. As we move, the fog lifts, allowing us to see the stations and making the whole process much easier.

After two hours, the ship's bow is in front of the ice blocks that are obstructing our mooring point. We push them away and soon we're back in our old position. It's all good!

July 16, 2020: Day 301

THE TEAMS ARE about to head onto the ice when someone spots a polar bear on the portside. The bear quickly moves closer, comes around the bow, discovers the thick orange power lines on our starboard side, and starts to play with them. When bears come right next to the ship, we scare them off by clanging a metal rod against the hull. From the bridge, I send colleagues armed with flare guns to the starboard side— where all our equipment is located—and to the main power distributor, where a big switch can shut down the entire network. The bear is rattled by the noise for a moment, but his curiosity gets the better of him, and he makes a beeline for our remote sensing instruments to starboard abeam. I radio our guards, tell them to get ready, and give the order to fire the flares when the bear starts pulling on one of the sensitive devices. A well-aimed flare detonates right next to him. He springs back and gallops away from the ship. We don't seem to have struck that much terror into him, though, and he heads straight for our mobile measuring sleds, which are a few hundred yards away on the ice.

The flares won't reach that far, so I need something else—the ship's horn. We can't sound the horn too often, or the bears will get used to it and realize that nothing bad actually happens when they hear the strange tooting. It's all about the timing. Through the binoculars, I watch the bear cautiously approach the station. Despite his curiosity, the bear is also stressed. He doesn't know what it is and can't tell whether it's harmless. If the horn is to be a successful deterrent, I need to exploit this stress. I wait for the moment when his curiosity overrides his respect for the unknown and he starts to sniff tentatively at the frame, then I press the button. It works! The bear jerks back from the device, runs away, and stares back in astonishment. He clearly thinks he

▶ Our Remote Sensing Site instruments, which examine the surface of the snow and ice from above. The large TROPOS tethered balloon can be seen in the background.

triggered the noise by touching the device, and has absolutely no desire to explore further, at least for now. He trots over to the distant station where we study two-year ice, which has drifted away on a floe. After the morning's excitement, he lies in the snow for a snooze. Every few minutes he blinks wearily and looks to see if we're still there. Otherwise he's pretty relaxed and obviously feels quite content in his cozy sleeping spot. The flares and horn don't appear to have had a lasting effect.

After power napping for an hour or so, he stretches out, clearly refreshed, and carries on exploring the strange new playground that has suddenly appeared in his home. I'd love to leave him in peace, but this bear is behaving differently from the many others that have visited recently. While they quickly lost interest and continued on their way, he remains fascinated by everything in our camp and is clearly growing more comfortable by the minute. That's not good. Our structures are no place for a bear and are distracting him from hunting for the seals he needs to survive. If he settles here, it will be detrimental to his health. We ready the helicopters.

The bear moves toward us with purpose, this time enjoying the thick power lines over by ROV City. We switch off the power immediately and proceed as arranged, firing a volley of flares in his direction. Bears shouldn't be chewing rubber, so we need to step in! As expected, we drive him away from the power lines, but not for long, so I order the helicopter to fly slowly toward the bear. The helicopter always works. We gradually drive the bear away from the ship and out of our downdraft.

Once the bear is 1.2 nautical miles away, the helicopter returns to the ship. With the weather as it is, the bear is beyond our range of vision. I modify the afternoon schedule so that nobody's working on the side of the ice floe facing the bear's last known position; this makes it easier to keep everyone safe. Then the teams head onto the ice. There are bite marks in the thick ROV City power line, so it needs to be replaced. If the bear doesn't show up overnight, we can work as normal tomorrow.

▶ HOW ARE POLAR BEARS COPING WITH CLIMATE CHANGE?

Polar bears live exclusively in the Arctic, in nineteen loosely separated populations off the Canadian and Norwegian coasts, the Siberian
islands, and even in the middle of the Arctic Basin. They are highly
adapted to the ice, where they hunt seals and find snow caves for
giving birth. This means they rely on the sea ice, which has become
thinner and younger in the past decades. The increasing amount of
thin one-year ice favors polar bears; seals need ice holes to breathe,
and these holes can only be found in thin sea ice. At the moment,
some bear populations in the central Arctic seem to benefit from
climate change, but this probably won't last forever. Data is scarce
because polar bears are much harder to count and monitor than their
land-based relatives. And if global warming causes the Arctic ice to
disappear completely in summer, the estimated 26,000 polar bears
will lose their habitat. If we don't stop the ice from shrinking, polar
bears will probably die out in a few decades.

▶ In summer, even polar bears have to cross open water. They jump short distances and swim across longer stretches.

July 17, 2020: Day 302

AT 4:00 AM the phone wakes me from my slumber. Our bear is back! As he rapidly approaches ROV City, I tell my colleagues to shut off the power. This bear seems to be attracted to power lines. Fittingly, the first thing he does is bite the orange cable. To prevent him from ingesting rubber particles, we fire the flare gun and drive him away from the power line. The bear canters away, then trots unhurriedly into the encroaching fog. When the fog lifts at around 6:30 AM, we find him less than half a mile from the ship, sleeping comfortably and occasionally blinking. He's still sleeping an hour later when the dense fog swallows him up again. When the fog clears later in the morning, there's no trace of the bear. The teams start working on the ice in the afternoon, with heightened security and an adapted schedule.

July 18, 2020: Day 303

THE BEAR RETURNS twice in the night, and he's there again in the morning. He strolls languidly along the edge of our floe and lies down for a snooze. He seems to feel comfortable in our presence; we feel anything but in his. We can't work on the ice this morning. We use the helicopter to drive the bear around two nautical miles away to a neighboring ice floe across a stretch of open water. We start working on the ice at 11:00 AM. We haven't seen the bear since.

There are now patches of open water all around us, with ice cover at just 60 to 80 percent. Our floe is drifting freely in the water, but we see no signs of instability.

Although the thawing season has only just begun, the Arctic sea ice map compiled each day based on satellite data is already showing huge ice-free areas in the Siberian Arctic. There's no ice at all in the Barents Sea (normal for this time of year), but even the Kara Sea, Laptev Sea, East Siberian Sea, and the Chukchi Sea are almost completely free of ice. It's startling. This side of the Arctic has never had such large ice-free areas in mid-July. Overall, this is the least sea ice the Arctic has had on any July 18 since records began. Only the Beaufort Sea has a little more ice than in some previous years.

The ice distribution is consistent with last winter's rapid Transpolar Drift, which pushed the ice from the Siberian Arctic to the Atlantic and Canadian sector faster than usual. Months later, we are still seeing the effects of last winter's unusual weather in the northern hemisphere, which accelerated the Transpolar Drift. Everything's connected. The remarkable wind we witnessed in winter is still affecting sea ice distribution—and this, in turn, is influencing the current weather. The wind's impact on the current ice distribution will have long-lasting repercussions for the ecosystem in and under the ice and will alter the flow of energy and substances between the ocean, ice, and atmosphere. Again, all of this will influence next winter's atmospheric circulation.

▶ Typical conditions on the ice in mid-July 2020.

The Arctic climate system is like an intricate timepiece; if one cog (or process) is disrupted, each aspect of the mechanism experiences long-lasting and unpredictable changes. Our mission is to understand how this timepiece works, how these processes interact. We aim to replicate the entire complex system in our computer climate models so that they can predict how changes in one area will affect the rest. We want to accurately map the entire mechanism. That's why we embarked on this expedition. And our efforts are going to pay off.

Throughout the day, we have planted a semicircle of flags of all the countries involved in MOSAiC. We spend the evening drinking mulled wine on the ice and barbecuing on board. We celebrate how well the expedition has gone so far. The mood is appropriately exuberant and we all party and dance. But by 1:00 AM we're tucked up in our cabins. We've been pushing ourselves to the limit day after day and now we need our sleep. Our team is fantastic: they work really hard, know how to relax and have a good time, but always remember their priorities.

July 21, 2020: Day 306

WE'VE EXTENDED OUR WORKING HOURS to optimize the limited time we have left on the ice floe. Two days ago, we did a night shift on the ice after our evening meal. Today, the Balloon Town and drone teams

▶ The *Polarstern* remains stable in the MOSAiC floe. As the upper layers melt in summer, trapped sediment collects on the surface and the ice floe slowly turns beige.

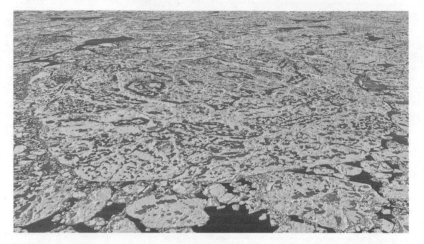

▶ The MOSAiC floe on July 22, 2020. The large melt ponds have now melted through and the accumulated sediment has created vast brown patches on the surface of the ice. The floe remains stable as the surrounding ice collapses.

started working at 4:00 AM. At 4:30 AM, the guard spotted a polar bear approaching through the fog around two hundred yards away. It was fine; we quickly evacuated from the ice and raised the gangway. The bear walked halfway around the ship to the portside, then moved off into the fog. When the fog lifted, the bear had left our line of sight. Our work continues.

July 22, 2020: Day 307

ANOTHER 4:00 AM START. It's wonderfully sunny and the sun burns down on the ice all day and night. Our ice floe has become an extraordinary lakescape. The melt ponds gleam an intense bluish green, their surfaces glittering in the sun. Their twisted shapes and rounded shores now cover more than half of the ice. We've had to build bridges across many of the meltwater channels. The strips of white ice between the countless lakes branch out in a never-ending maze—the height of summer in the central Arctic. We continue to drift south, closing in on the ice edge. We're almost at 80° north; we've passed Greenland's northeast tip and are drifting southwest in the Fram Strait at around eight nautical miles a day. The ice edge is just forty-five nautical miles to the southeast, but in the western part of the Fram Strait, an ice tongue stretches to the south along Greenland's east coast. And right now, it looks like that's where the drift is taking us. This will allow the floe to survive a little longer and carry us far to the south. But we'll reach the ice edge eventually, and then our floe will reach the end of its life cycle; it will disintegrate in the swell and waves, be carried out to sea, melt, and return to ocean water, from which it was formed almost two years ago off the Siberian coast.

▶ End of July, the height of summer on the ice. The floe has become a lakescape. We use bridges to access our research areas.

It will be important to predict exactly when the floe will break, so I take frequent helicopter flights to the ice edge and study the ice structures beneath me. There's a lot riding on the reach of the swell in these ice conditions and how much breakage it will cause. Initial, subtle changes to the ice structure become apparent at around fifteen nautical miles from the ice edge. The number of large ice floes declines, while areas of broken, ground-up ice gradually increase. This seems to be the point at which ice floes react to the ocean swell. At around ten nautical miles from the ice edge, the landscape changes completely and abruptly. There are no larger ice floes, and the surface of the ice consists entirely of tiny fragments. Stable ice floes break apart, unable to with-stand the swell. Naturally, these distances depend on the waves and swell on the open sea off the ice edge. If a storm whips up the water, the swell can penetrate much deeper into the ice and destroy floes farther from the ice edge. To avoid surprises, I now include wave forecasts for the North Atlantic in our daily weather briefings.

After completing today's explorations, we fly back over the vast ice fields in the stunning light of the low-lying sun. Summer has left its mark everywhere. From up high, the melt ponds that cover all the floes form an extraordinary pattern of interleaving blue-green bowls. The ponds are now beginning to melt through, leaving the ice floes with more holes than Swiss cheese. The floes are losing their integrity and most have already broken into several pieces. As we approach the *Polarstern*, our floe is a bastion of stability in an ailing ice field. Our research camp is as safe as ever, undeterred by the summer decay. We

▶ The ice floe, shortly before it disintegrates, in the typical thin fog of the Arctic summer.

also notice that while the flat areas of ice around us are now consistently covered in ponds, with barely any larger sections dry enough for our work, the ponds on our old, rugged ice floe cluster in the lower-lying areas and the higher ground remains dry. Thankfully, we had the foresight to set up our camp on the elevated sections. Only our L-shaped area of young ice looks exactly like the rest of the landscape; the only way to cross it is to don our survival suits and carefully wade across the ponds and their treacherous, unstable beds. But it's still firmly attached to the floe!

We land by the ship at 9:30 PM. I feel a close bond with our ice floe, which has carried us through the Arctic so loyally for almost ten months.

July 24, 2020: Day 309

FOR THE LAST eight days we have been visited by at least one polar bear every day, usually more. Since we sometimes have to deal with more than one creature at a time, we are constantly adapting our work. We've established routines for evacuating the whole or parts of the ice floe.

Otherwise, the usual cycles continue, alternating between opening ice and increasing pressure; however, the pressure isn't that great now and hasn't damaged the floe so far. When the ice opens, our floe floats freely in the water. The core hasn't cracked or deformed yet, but its edges are starting to erode. The thaw is in full swing. I figure that the smaller L-shaped area of one-year ice on one side of the floe could collapse at any time. We visit this area every day but don't leave any key equipment. The only permanent installation over there was ROV City, and that has now been moved to the two-year ice, our old fortress. We presume that the fortress—where most of the camp is located, and where the tall mountains are now melting and rounding—will remain stable for the time being. We're drifting south parallel to the ice edge, which is still almost forty nautical miles away.

▶ Ice formations at the edge of the MOSAiC floe.

I scout along our floe's shoreline to stern, then come around to the opposite side and return through the middle. Thin patches of fog. I'm alone most of the time but within view of my colleagues on the ice. The middle of the floe is like a huge desert, permeated by meandering wadis. It's beautiful. I stop on a tall pressure ridge on the other side of the floe. Pieces of ice float past on the open water, shimmering green. Bizarre, towering ice formations have developed on the edge of the floe, created by collisions with other ice floes.

We'll continue working here for as long as the ice allows. We've drawn up plans to quickly dismantle the camp if necessary. The *Tryoshnikov* will soon be leaving Bremerhaven with the team and crew for the fifth and final expedition phase. We'll meet the *Tryoshnikov* either here in the Fram Strait or near Spitsbergen. If the floe disintegrates before then, we'll continue with a temporary setup until the *Tryoshnikov* arrives. We'll then work on the move for a few days on surrounding ice floes. It's possible that we could drift worryingly close to the ice edge in the next few days and have to wrap things up. Nobody can say. When our floe finally comes to the end of its life, we will head north from Greenland's coast, but that's a matter for the final phase.

The People of the Arctic

THE ARCTIC MAY BE inhospitable, but it is inhabited—unlike the Antarctic, where humans are only ever temporary guests. Four million people live in the land regions bordering the Arctic Ocean; they include Indigenous peoples such as Inuit (approx. 150,000 people) from the Bering Strait to Greenland; Nenets (approx. 40,000) and Yakuts (approx. 330,000) in Siberia; Sámi (approx. 70,000) in Scandinavia and Russia; and Evenks (approx. 35,000), who are spread across Russia and Mongolia all the way to China, populating an area larger than Europe. (Then there are non-Indigenous Americans, Canadians, Scandinavians, Russians...) It's remarkable how the Arctic's Indigenous inhabitants have adapted their traditional ways of life to the harsh conditions. But climate change is having a significant impact on their ancestral homelands—and their lives.

When Heiko Maas was appointed German minister of foreign affairs in 2018, he declared that one of his priorities would be climate policy and the security issues raised by climate change. He wanted to take a preventive approach to conflicts sparked by climate change, rather than merely reacting to escalating situations. A positive and forward-thinking attitude. In August 2019, he decided to use the United Nations Security Council meeting in New York as an opportunity to take a detour to the Arctic on the way home, and he invited me to join him on his trip. We flew from New York in a government aircraft to Iqaluit in the south of Baffin Island, Canada. From there, we took a chartered flight north to Pond Inlet.

Iqaluit and Pond Inlet are small Inuit settlements in the Canadian territory of Nunavut, which has around 40,000 residents. Traditionally, Inuit live mainly by hunting marine and terrestrial mammals such as whales, seals, and polar bears. The soil is too barren for agriculture, so other foods are flown into towns like Pond Inlet; even the doctor only visits once every few months. As we flew into Pond Inlet, we could see the corrugated iron houses of its around 1,700 residents. We were greeted from the plane by Carey Elverum, manager of Sirmilik National Park. It was a beautiful, calm day, a balmy 14 °C (57°F), and the sun was shining. Children paddled in the little ponds that form in the summer on the permafrost soil. Around fifty years old, Carey said that he couldn't remember any summer as warm as this.

What he said matched with the weather station data. In 2019, a large portion of the Canadian Arctic experienced an unprecedented heat wave. In Alert, the most northerly permanently inhabited place on Earth (82° north, around six hundred miles north of Pond Inlet), temperatures reached a record 21°C (70°F). Fairbanks, Alaska's second-largest city, sweltered at over 30°C (86°F).

In Pond Inlet, we witnessed the despair on Carey's face. The weather may have been pleasant and allowed us a wonderful boat trip across Tasiujaq (formerly Eclipse Sound) to the Sermilik Glacier, but Carey explained why these conditions were so problematic. Normally, the ice in the sound would break open for just a few weeks in summer; the rest of the time, it serves as Inuit hunting territory. The ice allows them to travel on snowmobiles and connects the smaller surrounding settlements. If global warming extends the ice-free summer period, the hunting season will shrink and the settlements will be cut off from one another.

We sailed across the sound to the spot where, until a few decades ago, the Sermilik Glacier met and pushed out onto the water. There was no sign of the glacier; instead, we saw a vast expanse with loose piles of rocks surrounded by soft sediment that would be difficult to negotiate. This was the moraine landscape left behind by the glacier as it retreated. Brian, another member of the Inuit community who traveled with us, commented: "All this land! It used to be covered in glaciers. Now when I see all this open land, my eyes fill with tears—it looks so naked without the glaciers."

After trudging through the loose moraine for forty-five minutes, we reached the edge of the glacier. The warm day was thawing the ice; rivulets had formed all over the glacier, dribbling over the edge in tiny waterfalls. The dripping meltwater, murmuring rivulets, and burbling waterfalls created a striking soundscape. We could hear climate change in action! Obviously, it's normal for the lower section of a glacier to melt in summer. But these days, the ice that melts away each year far outweighs the ice added to the top of the glacier, and so it shrinks dramatically. Below us, a torrent of meltwater crashed out of the glacier gate on its way to the ocean, a striking illustration of ice mass being lost.

Heiko Maas was visibly affected by this manifestation of Arctic warming. He spent the whole trip soaking up every explanation of climate change and the links between the Arctic and the rest of the world. I could tell it was a topic close to his heart. He hadn't put the issue on his agenda out of pure political opportunism. When MOSAiC begins, he follows our progress closely; toward the end of the expedition, we hold a long phone call and recount our observations of the melting Arctic.

In Iqaluit we met Mary Ellen Thomas, who described her impressive family history. As recently as four decades ago, her extended family—who are scattered across Nunavut—would meet every year on July 1 (Canada Day) in Kuyait, which is over a hundred miles from Iqaluit. This was an opportunity to maintain family bonds. They traveled by dogsled, and later snowmobile, across the frozen fjords and coastlines. Back then, it seemed the ice would last forever. Twenty-five years later, large ice-free areas started to appear along their route, half of which was still passable. Today, Kuyait can only be reached by boat, and that's beyond the means of many ordinary people on Baffin Island. As a result, Mary Ellen's family is falling apart.

On our ice floe, we can feel what it's like for your home to melt beneath your feet. But we don't live here all the time. The people of the Arctic have no alternative. When the ice disappears, the unique landscape won't be the only thing that's lost. Millennia-old cultures and the way of life of millions of people are also at stake.

July 25, 2020: Day 310

OUR FLOE STARTED rotating yesterday! Depending on the ice field around us, it moves through more than 360 degrees per day at a breathtaking speed. We're still figuring out why. This rotation is pretty interesting in its own right. Since there's plenty of open water around us and we aren't colliding with thick ice, the rapid movement isn't destroying the ice floe, which is already fairly circular and can rotate freely. But when we next make harder contact with the surrounding ice, our unstable one-year ice will probably shear off.

We're prepared for that. The one-year ice typical for this area is slowly crumbling around us; all we see is broken ice. Only our floe remains stable, a tower of strength. We really did choose well!

July 26, 2020: Day 311

OUR CURRENT POSITION is just 79°45′ north. We left 80° north behind us yesterday morning and the drift has since carried us farther southwest, straight into the ice tongue in the western Fram Strait, which promises to extend the life of our ice floe. The open water of the central Fram Strait is just twenty-six nautical miles east of here, but as we drift southwest, the ice tongue extends far into the south. If we were to start drifting east, we could end up in the danger zone at the ice edge in just one day.

Each team has prepared an evacuation plan for their part of the camp. I'm confident that once I give the signal, we can get the whole camp back on board within a day. And that's the main thing now—to figure out how much longer we can work in safety.

I've seen the sea ice break at the ice edge plenty of times around Spitsbergen (in the 1990s, when it still had sea ice in winter). When the ice edge draws closer and the swell penetrates the ice, it doesn't break gradually—it positively shatters. What was once stable ice instantly transforms into a field of icy debris.

Previous expedition reports describe ice floes suddenly bursting with the impact of the swell. If you're on the ice when this happens, then you're in grave danger. When our floe begins to break, there won't be time to salvage our instruments. The whole camp would sink into the ocean.

On the other hand, our ice floe's final phase will be scientifically compelling, and we want to research for as long as we can. It's critical that we figure out when we need to dismantle the camp.

Decades ago, with the onset of every Spitsbergen spring, I would ask myself how much longer it would be safe to walk on the ice, and when it would shatter. People with more experience showed me how to spot the subtle yet visible signs of the swell in the ice, and I've since observed them many times. These signs usually indicate that the ice is about to break. You can spot the swell when the water gently rises and falls in the cracks and openings in the ice, or when the ice on either side of a crack occasionally lifts and lowers; this is often accompanied by the typical squeaking sound of floe edges rubbing together. When these warning signs appear, it's time to get off the ice.

And I have technical support on the ship. As of today, a swell monitor runs continuously on a screen in my cabin; I've collated it from the ship's data. Three lines show the rise and fall of the ship as well as her tilt along the longitudinal and transverse axes. I monitor the drift and the status of the ice edge day and night, and check the swell monitor every few hours, even overnight. For now, the swell is rising and falling by 0.4 to 0.8 inches at intervals of around eleven seconds and isn't yet visible in the ice. Our research continues.

July 27, 2020: Day 312

WE'RE STILL DRIFTING southwest in the ice tongue, the swell monitor shows no increase, and the ice is calm. It's going to be another good research day, accompanied by one of the polar bear visits that have become a common feature of our days. The teams all know that our hours on the floe are numbered, and we dig deep to take as many mea-surements as we can. Nobody expected that we would have so much time on this ice floe, and the fourth expedition phase has already been far more successful than we anticipated; when we returned to the floe, it had already drifted quite far south. Once again, the mood is excellent. We don't know how long we have left, so everyone enjoys their final days on the ice.

▶ In the second half of July, the camp is visited by polar bears almost daily, some-times by several bears a day.

July 28, 2020: Day 313

THIS EVENING I ANNOUNCE that, starting tomorrow morning, all infra-structure is to be removed from the ice floe.

Over the last few hours, our general drifting direction has switched from southwest to south, whereas the effects of the regular tidal cycles remain consistent. This means that after drifting parallel to the ice edge, we are now heading straight toward it. The latest satellite images also show that the ice edge has retreated rapidly toward us as the day has progressed.

When I made the decision at 5:00 PM, we were still around sev-enteen nautical miles from the line of 50 percent ice concentration;

now that it's evening, we are just fifteen nautical miles away. The core of our ice floe still has an average thickness of thirteen feet and there are no cracks or any other signs of instability (apart from a little erosion at the edges), but even this won't be able to withstand the swell around the ice edge. The swell has steadily increased throughout the day to two to nearly three inches in places. And if you look very closely, the swell is now visible in the ice. For me, this indicates that the floe is about to break. I have no choice but to give the order. Our floe has carried us for much longer than most of us imagined. It has provided us with a trusty and stable platform for our research until south of the eightieth parallel north. We have been able to document its life cycle from the Laptev Sea all the way to its demise at the ice edge in the Fram Strait.

The MOSAiC floe, and the ice camp, have reached the end of the line. We initiate our evacuation plans.

July 30, 2020: Day 315

WE DISMANTLED THE ICE CAMP yesterday and spent this morning collecting the remaining power lines, and not a moment too soon—while we were engrossed in our work, every team suddenly reported cracks in the ice. I went straight off to explore, one last trip across the ice floe. Away from the noise of the ship, we heard loud, muffled bangs like distant gunshots as the floe broke into more and more pieces. The signs of the swell are unmistakable now. The edges of the cracks rise and fall, crunching and squeaking. For the first time, it feels like we're standing on the sea. Over the last year, the static ice has often made it easy to forget about the ocean beneath us; now, we can see the ice undulating all the way to the horizon, the long waves of the open sea. Our home of the last year is dying to thundering booms. It's a strange feeling.

Descending fog creates a mystical feel as we return to the ship. We don't see her until we're almost there.

But we've seen enough. The ice floe is dying. Even now, it's nothing but little pieces that are only staying together because there's no wind to blow them apart. The cracks between them are only around four inches wide at the moment and can still be traversed.

We may feel sad, but our timing is perfect. We kept the whole research camp running until the very last day and returned everything to the ship in good order.

▶ The MOSAiC ice floe shatters on July 30, 2020, impacted by the swell and waves at the ice edge in the Fram Strait. The expedition stayed with the floe right to the end. The research camp had been dismantled the previous day.

July 31, 2020: Day 316

THERE ARE NOW barely any floe pieces larger than 150 feet. During the night they drifted in every possible direction; the floe has lost all its contours and there's nothing to indicate that until yesterday, this was a large, stable island of thirteen-foot-thick ice. The remains of the floe will now drift toward the ice edge and melt in the open sea. The death of an ice floe.

We use the ship to salvage the Met City hut from a piece of ice, then we raise a glass to the floe on the helideck. The MOSAiC ice floe is history. We turn on the main engines and sail away. Our work isn't finished yet.

▶ By July 31, 2020, the MOSAiC floe has completely fallen apart.

11

THE NORTH POLE — AND A NEW ICE FLOE

August 10, 2020: Day 326

WE'VE SPENT THE LAST FEW DAYS retrieving the equipment that we set up around the central research camp at the start of the expedition. We've sailed to every station in the network that has been sending regular GPS coordinates; in many cases, this is the only way to access the precious data and samples stored inside them. It might be more accurate to say that we've retrieved what was left of our equipment. Many of the ice floes have already shattered on the ice edge. Some pieces of equipment started off next to each other but have floated apart on tiny pieces of ice. Some were hidden in the fog; others ended up in the water and had to be salvaged from the ocean.

We've made good progress, but we're not quite finished. Any remaining equipment will be collected later by the *Akademik Tryosh-nikov*, which arrived today with our supplies.

▶ The *Akademik Tryoshnikov* and *Polarstern* meet by the ice edge in the Fram Strait for a supply run at the end of the fourth phase.

We've selected an area of open water, protected from the waves by large ice floes, in which to transfer fuel, cargo, and most of the team. We come alongside the *Tryoshnikov* and start to move the supplies while team members trade places between the ships. The whole area is buzzing with activity as our new colleagues are lifted over via mummy chair and crane. For a few days, the *Polarstern* is almost twice as full as usual. On a long expedition like MOSAiC, you need to ensure continuity between the various phases—even if each phase is covered by an entirely different team—so it's important to hold detailed discussions during the changeover.

August 16, 2020: Day 332

IT'S SUNDAY MORNING again. I'm thinking about goodbyes. The ship is quiet and glides through the water toward the North Pole. I listen to music and look back at the last few days. When the *Akademik Tryoshnikov* emerged from the fog on August 10, it dawned on us that we were about to go our separate ways. Around a dozen colleagues are staying on board with me for the fifth and final phase; the others are heading home on the *Tryoshnikov*. We've grown very close in the last three and a half months. The fourth phase is widely known as the "hugging phase." We've achieved incredible things and become great friends in the process.

▶ The *Akademik Tryoshnikov* returns home with most members of the fourth expedition phase. A few remain on the *Polarstern* for the fifth and final phase along with all the newcomers.

And then, on August 13, three days after the *Tryoshnikov* arrived, we faced the inevitable goodbyes. Most team members are leaving this distant, alien world for more familiar shores; others are staying till the end. We gathered on the helicopter deck one last time and there were hugs all around. Then the returnees were lifted onto the *Tryoshnikov* in the mummy chair. We completed the refueling process the following morning and then headed north. "Piano Man" played through the

▶ The *Polarstern* heads to the North Pole for the fifth expedition phase. First, she sailed through the floating floes west of Greenland; later, north of Greenland, she finds vast areas of open water.

loudspeakers (chosen specially by the captain) as we cast off and the *Polarstern* started moving with a loud toot of her horn. Both ships were a sea of waving hands, of laughing and crying faces.

Many new friendships—and a few couples—have formed in the previous weeks, and nobody knows whether these relationships will work when our friends return to the real world. We find ourselves wishing we'd spent more time with certain people, thinking of conversations we wish we'd had. The other ship grows smaller and smaller until our colleagues disappear, waving, into the fog. Eventually all we can see are indistinguishable dots of color. It feels like we've been abandoned. We'll miss them desperately.

But we also have new companions to greet! The team for the fifth phase includes many unfamiliar faces, as well as people we know from previous expeditions—and good friends who know exactly what we're going through after bidding farewell to our buddies.

We've been looking forward to seeing each other for months. When the two ships were close enough to make out the faces on the opposite deck, our sadness was alleviated by the joy of the new arrivals. I'm joined by a friend from an Antarctic expedition on the *Polarstern* in winter 2018/19. Back then, I left the *Polarstern* halfway through the expedition and flew home from the Neumayer Station. Our team had become very close, and I felt similar pangs as I watched the *Polarstern* cast off from the shelf ice edge, leaving Neumayer with her music blaring and horn tooting. A colleague and I stood on the ice, watching and waving as our friends sailed away, growing smaller and smaller.

My friend is lifted over in the mummy chair and we fall into each other's arms. It feels like two weeks since we last saw each other, not a year and a half; her presence is really helpful as I come to grips with the changeover.

▶ A film team follows the whole expedition. When Elise Droste cuts her birthday cake in the *Polarstern*'s cozy Red Salon, it becomes a media event.

Yes, the last few days have been stressful and emotional. The people who stayed on the ship were pretty exhausted and spent a lot of time dwelling on their thoughts.

But now we've reached the fifth and final phase and the newcomers are overflowing with enthusiasm, desperate to get started. They soon bring an end to our introspection. We're working on becoming a team again. Soon we won't remember who was here for the fourth phase and who's only just arrived. And that's just as it should be.

▶ WHO OWNS THE ARCTIC?

For centuries, various countries have hoped to unlock a route through the Arctic, only to find themselves blocked by a shield of ice. The United Nations Convention on the Law of the Sea now stipulates that the nations bordering the Arctic—Canada, Denmark, Finland, Iceland, Norway, Russia, Sweden, and the United States—are only permitted to use the area up to two hundred nautical miles off their coastlines (exclusive economic zones). But the Arctic's political status remains contentious. Canada claims rights to the Northwest Passage. Russia argues that because its continental shelf extends deep into the Arctic Ocean, it is entitled to vast areas of the Arctic right up to the North Pole—in 2007, it even planted a titanium Russian flag on the ocean floor to reinforce its claim. Former US president Donald Trump wanted to buy Greenland wholesale.

These disputes could intensify in the future. Now that global warming is melting the Arctic ice and making the region more accessible, old hopes will be given new life—hopes of profitable trade routes, of new fishing grounds, and of raw materials in the ocean floor, like oil, gas, and manganese.

After all, we have some big tasks ahead of us. It's time to add the final piece to our mosaic of observations, the only part of the Arctic sea ice cycle missing from our data: the initial freezing of the new ice, the "ice floe kindergarten," so to speak. It's mid-August and summer is drawing to a close. We can see the sun sinking lower in the sky, shining on the ice less intensely, and soon the summer thaw will give way to the winter freeze. We want to sail into the far north to record this process.

The daily ice maps we obtain from the satellites have shown us a potential route. In the last two weeks, the ice north of Greenland has torn wide open, and vast channel systems now extend almost to the North Pole. This is unusual. This region is normally filled with thick sea ice, multiyear ice in places, and everyone gives it a wide berth. The risk of getting stuck is too great. At first, I assume that these channel systems were caused by the wind driving the ice apart and tearing it open. If so, it would be tempting to risk the potential trap; if the channels remain open, we could cover the distance quickly and have plenty of time for measurements. We are in a hurry, after all—the freeze will begin any day now, and we want to set up camp on a new ice floe in the far north before it does. But if the wind changes direction, the trap will snap shut. The wind can close channels in an instant, and if the ice pressure grows, ships can end up stuck in the region for ages. The captain and I have been watching these channel systems for a while via the satellite data. Is this our highway to the north?

The alternative would be to sail far to the east of the ice edge and then head north, deep into the ice, off the coast of central Siberia. This would be a more conventional approach; the Transpolar Drift originates in the east and its ice floes are young and thin. But to get there, we would have to travel around the north of Russia's sovereign territory, for which we don't have a permit. We would have to travel deep in the ice to the east, and that would eat up a lot of time, even if it seems the more predictable choice.

The channel systems north of Greenland have continued to open for several days and aren't closing, even when the wind changes. What does it mean? Could these large areas of open water be more than just channels created by the wind? After a long night poring over the satellite maps in the captain's cabin, we decide to take the risk. If it pays off, it will be spectacular. It will save us heaps of time and take us straight across the North Pole. We set a course for the north, first along Greenland's east coast, then straight to the northwest, possibly into a trap.

And now here we are, at 87° north, just two hundred miles from the North Pole. We can hardly believe the journey turned out so well. Instead of narrow channels torn open by the wind, we see huge areas of open water, some extending all the way to the horizon. Any ice we find is crumbling, melted from above and below. It offers the ship little resistance. These aren't wind-made channels; this is large-scale melting. There is no trap. We race toward the pole at an almost continuous seven knots, the fastest the ship can go when there may be ice around. We're making such good time that we make multiple stops lasting several hours to study the water column in these extremely unusual ice conditions.

August 19, 2020: Day 335

THE NORTH POLE! We are standing on the Earth's axis, the point at which all longitudes and time zones meet, where compass points and times of day lose all meaning. For centuries, this place has fascinated humanity and fired explorers' imaginations. Many have lost their lives to the ice and cold in their attempts to reach the pole.

We arrive at the North Pole at 12:45 PM shipboard time, just six days since we were at the ice edge. A record time, and that's despite spending at least one day exploring the water column. As we approach, most of us congregate on the bridge, staring at the navigation computer, spellbound by the coordinates: 89.999 degrees north ... 89.9999 ... 89.99999 ... then the gyro compass starts to spin and the navigation system chirps with error messages. The compass can't work out our direction; the only way is south. We sound the horn to announce our arrival at the planet's most northerly point, and the watch officer brings the ship to a halt right on the pole.

▶ After sailing for just six days, the *Polarstern* reaches the North Pole on August 19, 2020.

What a sublime moment. It's a bit like midnight on New Year's Eve. We raise our glasses and . . . we're lost for words. What do you say at a time like this? Cries of "Happy North Pole!" erupt around the room as we celebrate this exceptional occasion, our spirits high.

The little GPS systems in our smartphones tell us that we can cross every longitude on board the *Polarstern* in just a few steps, and the ship's gyro compass continues to spin cheerfully. We gather on the helicopter deck for a quick photo before moving the ship to a small patch of open water slightly off the pole so that we can examine the water column again. And then we're on our way. There's no time to lose; the freeze is about to begin.

Once again, we're traveling largely through enclosed sea ice. The open water doesn't quite reach the North Pole. But there are smaller patches of open water here too, and all the ice has been eroded by the summer heat; well over half the ice is covered with melt ponds, some of them melted all the way through, and filled with holes. The ship encounters no real resistance. After crossing the pole, we travel swiftly south along the 105th meridian in the eastern Arctic. We are looking for a new floe somewhere along the route that the original MOSAiC floe drifted, around 120 to 180 miles from the North Pole. When we find a new floe, we'll set up our research camp again. We don't want to stay near the pole because the satellite data doesn't cover the ice conditions in the immediate North Pole region—most satellites no longer record the Earth's surface so close to the pole. Various satellites pass close by, but their orbits don't lead them directly over the pole.

Night at the Pole

WITH THE SPINNING GYRO COMPASS and beeping navigation system, this visit to the North Pole was more exhilarating than anything else. Things were a little different when I first traveled to the North Pole in January 2000.

I had just returned to Germany from a job at the NASA Jet Propulsion Laboratory in California. With my European and American colleagues, I then went on a major measurement campaign with several research aircraft in the Arctic to study the stratospheric ozone layer.

I was actually there to coordinate a network of around thirty ozone sounding stations in the northern latitudes, but when NASA planned a long research flight right over the North Pole in a DC-8 airplane, I was invited to join.

The flight to the pole went smoothly, and the instruments gathered data as planned. Obviously, it was pitch-black outside; it was polar night, after all. We could barely see the ice beneath us. The cabin lights were dimmed, and the control lamps and dozens of monitors glowed in the darkness. The seasoned operators sat at their instruments. There were no rows of seats like in a normal passenger plane. This flight was no different from the dozens of research flights I had taken in the past on various aircraft.

Well, maybe it was a little different. As we neared the pole, the tension mounted. Nobody knew how the plane's navigation system would react. And then, just as the display should have jumped to 90° north, the entire system bailed. We had no position data or any other navigation data. Not a great feeling when you're in the air. The pilots switched the controls to manual and kept an eye on the little handheld GPS systems they'd brought into the cockpit just in case. These systems were totally unfazed by the North Pole and conveyed our position throughout the flight. After a while, the plane's navigation system started up again and the flight continued as normal.

▶ The summer heat has eroded the North Pole ice to the point of breakage.

August 21, 2020: Day 337

WE HAVE SPENT one day and two nights traveling south on the east side of the pole. Last night, we crossed the eighty-eighth parallel north and entered the area covered by most satellites. I want us to find a new home as quickly as possible. Our approach is slightly different this time. At the start of the expedition, I had to select an ice floe that promised to support our research for an entire year and wouldn't fall apart as soon as the melt commenced the following summer. Now we need a floe that will allow us to research for around a month without collapsing. It needs to be thick enough to hold our ship stable; otherwise the first storm we encounter will simply push us through the ice.

Now that it's summer, the ice floes are so evenly covered by melt ponds that they barely show up on the radar satellite data. And there's dense fog pretty much everywhere we look, which means that other satellites capturing images within the visible spectrum don't get so much as a glimpse of the ice. The fog also prevents us from using the helicopter. Our only option is to look out the window, with some assistance from the ship's ice radar.

When I visit the bridge at 7:00 AM, watch officer Steffen Spielke has just abandoned his attempts to break through a broad, solid band of ice. He's rammed it several times but just can't crack it. Now he's sailing along it, searching for an opportunity to swerve around it. This hardly ever happens around here; up to now, we've moved through the ice with almost no resistance.

The solid ice now extends around a hundred yards along our entire portside and disappears into the fog on both sides. It's rugged and covered in deep-blue melt ponds, much bluer than the ponds on the surrounding flat ice. The color alone indicates that this is a massive piece of ice: these intensely blue ponds only occur on thick ice because less of the dark ocean shows through. The rugged piles of ice blocks are also extremely rounded, meaning that it is an older structure. Everything points to this being a broad shear zone from last winter in which the ice floes slid right on top of one another, cracking in the process. This is why the ice is so thick—it's the result of dynamic secondary ice growth. Our fruitless attempt to break through it shows that the pieces of ice have frozen together again after shearing, becoming solid. That can't possibly have happened in summer; this structure developed last winter.

There are vast areas of typical flat ice to the left and right of this ice band. The flat ice is less stable, but thick enough to work on; if we choose the strongest of the slightly elevated ridges, it will hold the infrastructure we need for the month.

Why continue searching? We've hit the jackpot. The thick ice band offers the stability the ship requires, and the surrounding areas seem perfect for our new study topic: the approaching freeze. There are even several easily accessible patches of open water nearby. I ask Steffen to stop the ship. Wow, that was quick! Instead of wandering aimlessly through the fog for days with almost no visibility, we have stumbled across a very promising ice floe.

I take a small team onto the ice for a closer look. The mummy chair ferries us over to the ice on the portside, and once again we find ourselves in a fascinating landscape. We search for a way through the labyrinth of ice blocks and discover new miniature mountain ranges behind every turn. Little valleys with deep-blue lakes, surrounded by mountains of ice. It's stunning. The flat ice stretches out on the other side of the mountains, permeated by typical greenish melt ponds. Here too, there are slightly elevated, flat ridges where we can construct our research cities. We drill into the thick ice between the ridges and see that the stability of the ice is consistent enough for us to work safely.

We extend our excursion to find a suitable mooring point for the *Polarstern* and leave a surface marker buoy so that we can find it again in the fog. We plant flags in the ice along the approach route to the mooring point.

▶ On August 21, 2020, we find our MOSAiC ice floe, version 2.0—the home of the next expedition phase. It features a whole range of icy landscapes.

In front of the icy mountains, there's a flat shoulder that we can use as our logistics zone; the crane will deposit heavy equipment here to be moved elsewhere. There's also a path across the mountains. With a little work, we can build a track for the snowmobiles—our very own mountain pass.

We quickly return to the ship, and the captain and I discuss how to maneuver it into the ice. We sail in a wide arc to the spot we've marked and, as before, attach our starboard side to the solid ice—it's easier to operate the crane on that side. By the afternoon we have assumed our final position, and the team leaders join me on the ice to sketch out our new camp. In the evening, we retire to the Blue Salon to turn our initial ideas into specific plans for tomorrow.

August 23, 2020: Day 339

YESTERDAY WE BEGAN to set things up. We laid the power lines on the ice floe and started up some of the instruments, but then a crack appeared across the icy mountains to the stern. It ran past our planned camp—so our setup can remain the same, at least—but it separated our loading zone and the point where the power line runs from the ship to the ice from the rest of the camp. And it meant that the gangway at the back of the ship touched down on the wrong side of the crack. I called a halt to the construction work and monitored the situation. By the afternoon, the crack had opened into a serious lead, and I decided that we needed to move around one ship's length forward before setting up anything else. We did that this morning, and now the whole ship is in the stable ice. Our mountain pass is very close to the crack, but it's still the best way to access the ice on the other side behind the ship. We leave the power line for the Remote Sensing Site, Met City, and Ocean City where it is for the moment and move the cable for ROV City (which is in front of the ship) to a new path across the mountains; if the stern access point is cut off by an approaching polar bear, we'll have an alternative route back to the ship. I always make sure we have more than one option. I'm still looking for a route across the mountains in front of the ship.

At midday, we resume all construction work.

August 25, 2020: Day 341

WE SET UP THE CAMP in record time and are taking as many measurements as we can. We've obviously learned a great deal since we built our first camp almost a year ago. We're much faster and have improved equipment mobility. In short, we really know what we're doing.

Today I join the drone team at the drone airfield, which is far ahead of the ship but still behind ROV City and the area we've kept free for ROV measurements. During the night, a crack opened between ROV City and the airfield that's too wide to cross on foot or with a simple bridge of Nansen sleds or planks. I paddle across with one of our pontoons and we set up a cable ferry across the crack.

My task is to watch for polar bears on the other side of the crack. Thanks to the pandemic, not enough of the scientists have been trained to use our weapons, so now we have a guard shortage; I step in most days. While Roberta Pirazzini and Henna-Reetta Hannula prepare their drones for flight, I find a tall ridge and patrol up and down.

▶ At first, we use kayaks and pontoons to cross cracks and leads in the ice. Later, we set up cable ferries over the water.

I like to keep moving while watching for polar bears. It changes your perspective and makes it easier to spot sleeping bears behind blocks of ice (bears like long sleeps). It also helps you to maintain your concentration. Every so often I stop, rest my binoculars on the probing rod I always carry on unfamiliar ice, and scan the area. No bears to be seen.

The atmosphere on the floe is extraordinary. Again there's a thin layer of fog on the ice, the *Polarstern* hazy in the background. The sun shines through the fog and, as usual, an impressive fog bow swoops through the sky. The melt ponds shimmer in turquoise. Most of the ponds are now covered by a delicate layer of ice that enhances their pastel tones. This ice won't last long—the freeze hasn't yet begun in earnest, but it's announcing its presence.

My colleagues fly the drone over the ice to measure the radiation coming from above. But Henna loses control of the drone almost as soon as she launches it. Flying drones is extremely challenging at these latitudes. Their magnetic compasses don't work so close to the North Pole, so the automatic steering fails. Operating the drone manually, Henna quickly realizes that something isn't right. She skillfully makes an emergency landing on a distant, frozen melt pond.

▶ The light on the ice in late August 2020.

As soon as we retrieve the drone, we can see why it wasn't working. The front edges of the rotor blades are totally frozen. The ice has altered the blade profile completely, creating an upper front rim that intercepted the current. The drone lost its uplift, and if Henna hadn't acted so quickly, it would have crashed and been destroyed. This thin fog is treacherous for all flying machines. It's made up of minuscule, supercooled water droplets that freeze as soon as they meet a surface. Icing has caused numerous crashes throughout aviation history and will remain with us throughout the last expedition phase, testing the patience of the helicopter and drone pilots as they wait for conditions to improve.

▶ The icy landscape near the North Pole.

▶ October 2, 2019: The *Polarstern* meets its escort ship, the *Akademik Fedorov*, in a thin-ice area.

▶ Two polar bears visit the *Akademik Fedorov* and *Polarstern* at their meeting point.

▶ Icebergs, bizarrely distorted by mirages, in front of the shelf ice edge near Atka Bay in the Antarctic.

▶ The icebergs in the image at left do not exist, and neither does the bay in the ice behind the aircraft (below). They are both mirages near the German Neumayer III Antarctic station. In reality, the flat shelf ice reaches all the way to the horizon.

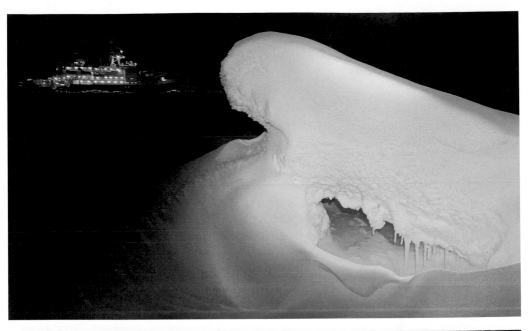

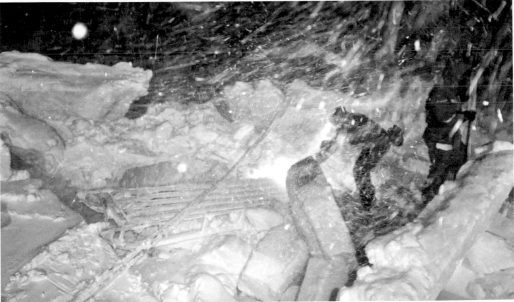

▶ November 2019: Ice formation in the fortress behind Balloon Town.

▶ December 2019: A salvage operation in the midst of a storm as a huge pressure ridge develops and threatens to devour crucial equipment.

▷ December 18, 2019: The *Kapitan Dranitsyn* leaves
the *Polarstern* and begins the long journey back
through the ice.

▷ June 28, 2020: The drone team returns to the ship
after a successful day on the ice floe. The great
summer thaw has begun.

▷ July 2020: A polar bear inspects our camp.

▷ July 13, 2020: The two tethered balloons float above the summery ice floe.

▷ August 2020: In no time at all, the MOSAiC 2.0 research camp is up and running.

▷ In the Arctic summer, the thin, shallow fog creates fascinating light displays and frequent, striking fog bows opposite the sun.

▷ September 9, 2020: Halos appear around the low-lying sun.

▷ Early September 2020: Working on the ice.
  As the late summer sun draws closer to the
  horizon, the colors intensify.

▷ September 20, 2020: A contemplative farewell
  to the ice floe in our final minutes on the ice.

12

GOING HOME

September 2, 2020: Day 349

WE'RE SO CLOSE to the North Pole that the sun moves around us constantly, its distance from the horizon almost unchanging. But as the weeks go on, it sinks lower and lower and draws closer to the horizon. At the end of September, it will drop below the horizon, and then polar night will recommence.

The sun's lower position is noticeable already. The light is becoming warmer and yellower, the colors more intense. This is an incredibly beautiful time to be on the ice. With its mountains, valleys, and flat lakescapes, our ice floe is pretty as only Arctic ice can be. Some of the team refer to it affectionately as our "beautifloe."

The camp is in full working order and all teams are working intensively. The freeze is imminent, and then the landscape will abruptly change. The melt ponds will freeze over and be covered in snow, and the summer lakeland scenery will suddenly revert to the frozen landscape from the start of our expedition. We want to record all the details of this rapid transformation.

▶ Traveling across the MOSAiC 2.0 ice floe in early September 2020.

▶ In early September, the sun moves noticeably closer to the horizon as it continues its eternal march. The light is warmer and more yellow. In just three weeks, the
sun will disappear behind the horizon and polar night will begin.

September 8, 2020: Day 355

YESTERDAY AFTERNOON WE BEGAN another of our frequent and intensive measurement cycles. This time, we'll be working on the ice for

thirty-six hours solid! Most of the instruments in Met City and the Remote Sensing Site operate around the clock anyway. But my colleagues in Ocean City will continuously lower a probe into the water to measure the chlorophyll, salt content, and temperature and detect where chlorophyll, salt, and water mix in the water column. The Eco Lodge tents also have a new home and are always occupied. In addition, our research balloons launch from the ship every three hours, more frequently than before.

We've arranged this measurement cycle to coincide with a change in the weather predicted by our meteorologist. First the wind grew stronger and stronger, then it started to snow. The wind and snow culminated last night; then, in the early morning, the cloud cover parted to reveal the gleaming sun. The stunning view more than compensated my colleagues for their night shift on the ice.

▶ MOSAiC 2.0 during an intensive round-the-clock measurement cycle.

The temperatures have also fallen. Our ice floe changed today and is now shrouded in a thick layer of snow where previously we walked across sheer ice. And the melt ponds have finally frozen; before, the occasional thin layers of ice would always melt again. This is the long-awaited "freeze-up" so crucial for our research—the final transition from the summer thaw to the winter freeze—and exactly what we expected from our meteorologist's forecast.

Things couldn't have gone any better. We have completed an entire year of measurements covering every phase of the annual sea ice cycle. We have witnessed one full heartbeat of the ice. And now we find ourselves back in the frozen landscape we encountered at the start of the expedition. The circle has closed.

September 9, 2020: Day 356

THE LIGHT EVOLVES every day, and it's fascinating. Each day the sun moves slightly closer to the horizon, the light becomes a little warmer, and now the yellow tones of the last few days are slowly turning a warm yellow-orange.

Today the sky plays host to an extraordinary spectacle. A striking pattern of geometric lines adorns the sky around the sun. We can clearly see the 22° and 46° halos, two huge concentric rings around the sun that intensify to form four sun dogs to its left and right, a light pillar rising from the sun, and an upper tangent arc on the inner halo. These lines of light are created when sunlight is reflected and refracted on the ice crystals floating in the air. They are a frequent and impressive sight in the polar regions.

▶ Halos appear around the low-lying sun on September 9, 2020.

September 12, 2020: Day 359

WE'VE CHOSEN OUR departure date: in eight days' time, we'll pack up our tents and commence the long journey out of the ice. The end of MOSAiC suddenly feels very real, and our thoughts return to the rest of the world.

For months we have been living in a little COVID-free community, isolated deep in the Arctic. We've practically forgotten about the virus. We don't need to social distance and can party without rules or concerns—a virus-free paradise in a world that has totally changed over the MOSAiC year. We can't imagine what life will be like when we get home. We receive daily news summaries from the mainland, but we barely take it in—the rest of the planet seems so far removed from our reality. Today, my friend Laura writes from Germany that European infection rates are rising again, and many countries are considering stricter measures to get the pandemic under control. We can't push it to the back of our minds any longer; our expedition is drawing to a close and we will have to deal with the pandemic. Some team members suggest that we stay in the ice until the pandemic's over. A nice idea, but unworkable.

▶ The *Polarstern* at the MOSAiC ice floe on the day before departure.

The Arctic—a Changing World

OVER OUR YEARLONG EXPEDITION, we have come to know the Arctic better than ever before. We've felt its heartbeat and stayed with our ice floe through every phase of its life. The new knowledge we have acquired is unprecedented, and it's not done yet; our samples and data will be studied in labs and computers for years to come.

Some things we do know: In the summer of 2019 and 2020, the ice retreated faster than ever before. The ice spread only half as far as it did decades ago and was only half as thick as in Nansen's day. In the winter, the MOSAiC expedition recorded temperatures that were almost consistently 10°C (18°F) higher than those recorded during the Fram expedition around 125 years ago. These are clear examples of just how quickly the Arctic and its climate are changing. Any measurements we don't take today won't be possible in a few years because the Arctic will be a different world by then.

MOSAiC allows us to understand this change. By conducting a yearlong expedition, we can better analyze the processes that cause the Arctic to heat faster than any other region on the planet. Complex mechanisms create close connections between the Arctic's atmosphere, snow, sea ice, ocean, ecosystem, and biogeochemistry. These processes don't just intensify climate change, but are also altered themselves. MOSAiC recorded over one hundred complex climate parameters all year round that will help us to understand these processes. We can now replicate them in our climate models to better estimate the impact that specific quantities of greenhouse gas emissions will have on the climate in the Arctic and worldwide—crucial information for political and societal decision makers who use science to guide their future climate-protection measures.

Never before have so many complex instruments worked simultaneously in the central Arctic as they have on our ice floe. We recorded the Arctic's entire heat budget and measured how energy spreads in the form of light and thermal radiation, and how it is carried by the tiniest amounts of water and air turbulence. We measured how heat from the ocean is conducted through the ice and snow and warms the surface. We measured how the surface cools by emitting thermal radiation, and how it is warmed by thermal radiation from the atmosphere, clouds, and aerosols. We recorded precisely how ocean eddies carry heat up to the ice from the depths of the ocean and how heat spreads through the atmosphere via air turbulence. Combined, these factors determine the temperature in the Arctic climate system. Changes to this flow of energy increase Arctic warming; now that we understand this more fully, we can better incorporate it into our climate models.

The *Polarstern* voyage lasted over twelve months. In winter, the atmosphere in the northern hemisphere was dominated by an unusually pronounced wind pattern, with a westerly jet stream around the Arctic stronger than any since records began in 1950. This wind pattern caused a rapid Transpolar Drift from Siberia to the Atlantic via the North Pole, and we were right at its center.

Everything in the atmosphere affects the amount of light and thermal radiation it sends to the surface of the ice. We recorded how the clouds interact with the sunlight, the heat they emit, and, in particular, how this changes depending on the cloud characteristics. We also measured how the tiniest aerosol particles influence clouds; up to now, these cloud processes were one of the greatest unknowns in the Arctic climate system. We now know what proportion of water droplets in clouds freeze to become ice crystals, the conditions in which this happens, and how this determines the clouds' impact on light and thermal radiation. We have now improved the way our climate models replicate clouds and their climatic influence.

We worked on the thin layer of ice and snow that separates the atmosphere from the ocean. In winter, when the atmosphere is much colder than the ocean, this insulating layer slows the cooling of the water and thus the freezing of new ice. We now know how well this insulation works and how it is affected by cracks forming in the ice. We understand better how snow spreads across the ice, creating another insulating layer, and how it is redistributed by the wind. We measured the mechanical properties of the ice and gained more knowledge of how the thinner ice moves, influenced by the wind and ocean currents. We watched—and felt—pressure ridges over ten feet thick forming in a matter of hours; we also saw how quickly the sea ice can tear, even in winter and spring, to form leads several hundred yards wide. And we studied in detail how, as the summer thaw begins, a remarkable layer of fresh water forms beneath the ice, almost like lakes—another barrier for the exchange of energy and gases between the water and atmosphere, and for the nutrients required by the lifeforms in and below the ice.

The Arctic ice is a unique habitat. We spent the entire year researching the Arctic ecosystem on, under, and even inside the ice: polar bears, seals, Arctic foxes, fish, and countless microorganisms have shown us how the ecosystem works, even in the dark of polar night. The microorganisms in and under the ice are the foundation of this food web. MOSAiC's detailed research has improved our understanding of how the sensitive ecosystem works in this extreme region and the changes its lifeforms are experiencing as a result of climate change.

Algae in the Arctic Ocean produce dimethyl sulfide, a gas that forms aerosols in the atmosphere and influences cloud characteristics. In turn, aerosols and clouds affect the fundamental flow of energy in the Arctic. We studied these processes, measured the dimethyl sulfide in the water and air, and observed how it escapes into the air through cracks in the ice. We also watched how this impacts aerosols and clouds. We now have a better understanding of how life influences the climate. We also measured how the two most important greenhouse gases—carbon dioxide and methane—are absorbed and emitted by the ice, and how this happens on the surface of the water in ice cracks. This means we can now better determine the Arctic's role in the atmosphere's global greenhouse gas budget.

And so the circle closes. From the outset, MOSAiC was designed to comprehensively study every aspect of the Arctic. It wasn't always easy, but we did it. This year will occupy us for a long time to come, both personally and scientifically. Perhaps for even longer than we can currently foresee.

September 20, 2020: Day 367

IT'S TIME TO SAY GOODBYE to the Arctic ice. After exactly one year on the expedition, we are beginning the slow journey home. It's been a long year, and for the first time I feel a certain relief. Having been responsible for the safety of every person on the ice, every single day, I gradually relinquish my role. We will arrive in Bremerhaven on October 12.

In the morning, we dismantle what remains of the camp and take our final measurements on the ice. We have to stop working around midday; the polar bear who's spent the last few days watching us from a safe distance starts to move closer. At 3:00 PM, the bear stops by the lead behind the ship, far enough away that we can go onto the ice floe and commemorate the end of our expedition.

We take one last group photo. Suddenly we're all pelting each other with snow, laughing and cavorting, reluctant to let go of what will probably be our last time on the ice. Then the kitchen team bring out the mulled wine and we enjoy our final two hours. We wander across the tall ridges and watch the sun moving along the horizon; the two are almost touching now. We see the polar bear hunting on the lead's thin ice, and then two lively seals pop up behind the ship, waving goodbye with their flippers. It's a magical moment.

As our departure time approaches, so does the bear; clearly, it's time for us to board the ship. Once everyone's on board, I walk up the gangway. We lift it on deck with the crane, start the engines, and head south in the deep orange-red sunlight, out of the ice, back to a distant world utterly transformed by the pandemic. We can't even imagine what awaits us.

▶ MOSAiC IN NUMBERS

- The *Polarstern* drifted over nine hundred miles from the nearest human settlement.
- The ship drifted for three hundred days with the first MOSAiC floe and another thirty days with the second.
- Her zigzag route with the ice covered over two thousand miles— about twelve hundred miles as the crow flies.
- The lowest temperature was –42.3°C (–44.1°F). Including the wind chill, the lowest perceived temperature was –65°C (–85°F).
- 1,553 research balloons were launched during the expedition, the highest reaching 119,022 feet. The deepest ocean measurement was taken at 14,098 feet.
- 135 terabytes of data were collected on the original ice floe, plus countless ice and water samples and aerosol sampling filters.
- More than sixty polar bears were sighted around the *Polarstern*.
- The expedition participants came from thirty-seven nations.
- There have been zero expeditions like MOSAiC.

For more information on the MOSAiC expedition, including photos, interactive graphics, and the latest updates on the mission's scientific research, visit mosaic-expedition.org.

▶ The final days on the ice before the MOSAiC expedition departs on
September 20, 2020.

▶ The *Polarstern* on the long journey out of the ice, already in brighter, more
southerly latitudes around the ice edge.

EPILOGUE

AFTER A THREE-WEEK JOURNEY, we sailed into Bremerhaven on October 12, 2020, 389 days after the expedition began. Returning to civilization proved overwhelming. Our arrival was reported live on TV and we were escorted into the harbor by a whole fleet of ships and boats. Our minds were still in the ice, but the hubbub around us quickly brought us back to the real world. It was a very warm welcome.

All expedition members returned from the Arctic safe and sound. One person suffered a simple leg fracture in the first phase; he traveled back on the *Akademik Fedorov,* our initial escort ship, and recovered in time to rejoin the third phase. Many who were involved in the first three phases (the coldest) experienced superficial frostbite to their faces, but we all soon healed. The frostbitten finger from the first phase was also cured, and many minor injuries were treated on board. I was incredibly relieved when we entered the harbor, our mission successful.

We're back on land now, but we're not the same. We have seen what's happening in the Arctic with our own eyes. As well as our data, we carry sensations and memories that will stay with us forever. This year has changed us, and it will change the many branches of science that deal with Arctic climate change.

But how else will the expedition make an impact? Will it affect how we treat our planet? Is there still a way to save the Arctic's year-round ice? If you'd seen the crumbling, melted ice at the North Pole in summer 2020, you might have your doubts.

The Earth's climate system has various tipping points that, if triggered by global warming, will lead to irreversible change.

Imagine a marble in a mountainous landscape with flat and deep valleys. Right now, the marble is happily rolling around one of the valleys; its movements represent our weather, its average range our climate. If we influence the system by pushing the marble slightly to one side, it will move around the valley differently—that is, our weather and climate will change. If we leave the marble alone, it will return to its old movement pattern—the change will be reversed.

But if we disrupt the system too much—if we push the marble over a mountain pass into a neighboring valley—then we change the climate forever. Even if we regret our actions and never touch the marble again, it will never return to its former position. These mountain passes are the climate system's tipping points.

Soon, we will cross the mountain pass that will erase the summer sea ice from the Arctic. When we do, we will find ourselves on a steep slope with no way back—no way to prevent the Arctic from losing its summer ice. And this is just the first tipping point that we will cross as global warming increases. At a global warming level of just 1.5°C (2.7°F), the risk of triggering further tipping points increases. The Greenland and West Antarctic Ice Sheets may become unstable and eventually disappear, the alpine glaciers are under threat, and we no longer know whether the coral reefs can be saved.

If global warming exceeds 2°C (3.6°F), then further tipping points lie in wait: the Amazon rainforest could vanish, thermohaline circulation (ocean currents) could fundamentally change, and the very existence of boreal forests might be jeopardized.

If global warming increases even further, we may find ourselves saying goodbye to the East Antarctic Ice Sheet, the Siberian and North American permafrost, and even the Arctic's winter ice.

We don't know exactly where many of these tipping points are, or how much global warming is required to set them off. Beyond the 1.5°C boundary there lurks a minefield, and we have no idea where the mines are hidden.

One thing we do know is that the disappearance of the Arctic's summer sea ice is one of the first landmines in that field. I wonder whether we might have stepped on this already and are now watching the explosion begin.

Do we really want to walk into this minefield with our eyes covered, or to send in our children? That is not a good or responsible way to behave. We need to stop disturbing the marble. We need to drastically and rapidly reduce our greenhouse gas emissions, carbon dioxide (CO_2) first and foremost.

To avoid entering the minefield—to restrict global warming to 1.5°C—we need to reduce global emissions to net zero by the middle of the century. This means releasing no more greenhouse gases into the atmosphere while also extracting them from the atmosphere. It's questionable whether we can still limit global warming to the 1.5°C threshold. But every tenth of a degree of warming beyond this limit increases the risk of serious global consequences. We must all take meaningful action so that future generations are not forced into the minefield, or at least not too far.

To do this, we must set realistic goals and introduce measures that will help us achieve them. To be successful, concepts for reducing CO_2 emissions must fulfill two basic conditions.

First, they must be effective. This seems so obvious that it's barely worth mentioning, and yet none of the climate-protection measures taken over the last few years actually meet this requirement. These measures alone will not allow us to achieve our goals.

Second, they must have majority appeal. In a democracy, lasting change can only be achieved with the support of the social majority.

So we have a dilemma. What if these two conditions contradict one another? Right now, this appears to be precisely what is happening: a broad majority of society agrees that "more must be done to combat climate change," but very few specific, effective suggestions manage to garner majority support. Political figures (who always need to consider whether their proposals appeal to the majority) often resort to placebo measures to show that they are acting on this increasingly important topic.

Most of us believe that we have to "do something" about climate change. Some people want to capitalize on this and commit society to restricting democracy, so that the question of "what to do" can be solved by forcing the majority to comply with measures designed and declared by a minority. But such an approach will never be successful in the long run, and it would undermine one of the greatest achievements of history—democracy itself. I really worry when people like Roger Hallam, cofounder of Extinction Rebellion, make comments such as "when a society engages in morally corrupt activities, democracy is irrelevant."[4] Statements like this discredit reasonable demands for climate protection and undermine the pursuit of social majorities. In doing so, they seriously compromise climate protection.

Even the most well-meaning appeals have the potential for societal backlash. This isn't about gaining the vocal support of people who already advocate drastic measures to protect the climate. And it's not helpful to focus on a few diehards who'll never be won over to the cause. The main thing is to convince all the people who are sitting on the fence. You can't protect the climate if your exorbitant demands scare them away. Polarization won't get us anywhere. We need to find solutions that most people can accept.

Climate protection requires the support of a social majority who understand why it is necessary. This is the only way to ensure long-lasting, cross-generational success.

Ultimately, there is only one way to mitigate climate change, and it's not easy: to develop a concept that fulfills both conditions in equal measure, and then to promote this concept to majorities patiently and relentlessly.

I am absolutely convinced that our society has the ability to reason. With such clear scientific foundations, anyone willing to explore the issue will recognize the urgent need for action.

The last two years have seen a dramatic increase in the number of people willing to engage with the topic. There has never been a better time to gain majority support for specific and effective climate-protection measures. We need to seize this opportunity, rather than tainting the issue further with ideological trench warfare.

To attract majority support, measures must meet the following three requirements:

1. They must be fair and balanced. They need to treat all CO_2 emissions equally, across all sectors, with no ideological bias. The best solution here is standardized pricing for CO_2 emissions, regardless of where and how they are generated. This could take the form of a tax, certificate trading, or another mechanism yet to be developed.

2. They must be supplemented with regulations that prevent displacement abroad. Otherwise, CO_2-intensive processes will just be relocated out of those countries that implement a carbon-pricing scheme. This is perhaps the most difficult thing to carry off; obvious measures such as CO_2 pricing for imports could conflict with current World Trade Organization trade rules, which are highly valuable in themselves. World trade experts need to put this topic on their agenda and look for solutions. This process is already underway.

3. They must be designed in such a way that 100 percent of revenue generated by these measures is returned to the people. Majority support would be precarious if the state appeared to be filling its coffers in the guise of climate policy. Plus, without such returned revenue, there's a very real danger of social imbalance; people on lower incomes are less able to absorb increases in energy costs (e.g., heat, electricity, fuel) than those on higher incomes. Actions must be taken to counteract this issue.

However, if CO_2 tax revenues (for example) were evenly distributed throughout the population, people on lower incomes would actually feel the benefit more—as a rule, they tend to use less energy than people in higher income brackets. The amount reimbursed would exceed any additional costs generated by increased energy prices, while still rewarding everyone for reducing CO_2 emissions. If this sort of mechanism were implemented and explained in a suitable and comprehensible way, I'm sure it wouldn't take long to achieve majority support.

Every successful climate-protection concept eventually develops incentive effects, and that's a good thing. These effects must not be ruined by exemptions or compensation regulations for the industry sectors and social groups who will inevitably start to complain that they are feeling the pinch the most. However, we also can't allow trust in predictable, reliable, and responsible government action to be undermined.

Here's an example: If you live a long way from your workplace, in an area where rents are lower, then rising mobility costs will make life more expensive (compared to living in the city and paying more rent but without the long commute). This is as it should be, because a longer commute generates more CO_2. The thing is, choosing a place to live is a long-term decision, so the state must ensure that conditions won't change abruptly, that people won't suddenly have the rug pulled out from under them; being able to rely on the state is another major social asset that we mustn't relinquish easily.

New climate-protection regulations must include solutions to help people transition to the new, reliable conditions, even if this seems to contradict climate protection in the short term. This will demonstrate that the state is acting fairly, responsibly, and with balance, which is essential in obtaining majority support for climate measures. After all, without a majority, nothing will happen.

Humanity has solved other global environmental problems through collective action. The ozone layer is on the road to recovery because every country on Earth signed the Montreal Protocol (and its subsequent addenda), which committed them to stop producing substances that were destroying the ozone layer. Despite the many conflicts of interest between different countries, we are capable of coming together to act responsibly. We are not a rabbit sitting in front of a snake, waiting passively to see what happens.

And if some countries don't play their part, then the others must lead the way. Just like protecting the ozone layer, protecting the climate is a task that will last for generations. Admittedly, the ozone issue was much easier to solve, but it shows that, in principle, we are capable of global action.

Unfortunately, our world is increasingly shaped by clashing national interests. This has to change, and we are the ones to effect that change. Multilateralism—the counterpart to nationalism—must prevail, along with international collaboration and joint responsibility for our planet. Otherwise, we have no hope of tackling the enormous challenge that climate change presents.

The Arctic sea ice isn't just an important part of the global climate system; it's integral to ancient cultures and the foundation of life for many Indigenous communities. And it's fascinatingly, extraordinarily beautiful. We should do everything we can to preserve it for future generations.

ACKNOWLEDGMENTS

THE MOSAiC EXPEDITION would not have been possible without the tireless efforts of hundreds of people, on board and on land, both during the expedition and long before it began.

I am especially grateful to Klaus Dethloff, MOSAiC's tenacious father and co-coordinator. He came up with the idea over a decade ago and developed and pursued it for many years. Without Klaus Dethloff, there would have been no MOSAiC. I would also particularly like to thank Matthew Shupe, another MOSAiC co-coordinator, for his early and visionary role in the MOSAiC program and his unceasing efforts before and during the expedition. I'm also extremely thankful to Anja Sommerfeld, MOSAiC's project manager, whose amazing commitment and incredible energy helped bring the project to life.

Thank you to Uwe Nixdorf, AWI's vice director and head of logistics and the mastermind behind MOSAiC's logistics concept. The expedition would not have been possible without the AWI's exceptional logistics department. I thank the logistics team, particularly Marius Hirsekorn, Verena Mohaupt, Bjela König, Eberhard Kohlberg, Tim Heitland, Dirk Mengedoht, Nina Machner, and many, many others. I am grateful to Michael Thurmann and everyone at the F. Laeisz shipping company for their excellent teamwork, their outstanding support as we planned and executed the expedition, and their expert handling of MOSAiC's many complex procedures.

Karin Lochte and Antje Boetius, former and current AWI directors respectively, deserve extra-special thanks for their tireless and staunch support for MOSAiC and their efforts to advance and implement the expedition plans. During her time at the AWI, Karin Lochte played a key role in building the international momentum that allowed the expedition to happen in the first place. Antje Boetius has always supported MOSAiC in every possible way. Her amazing commitment was pivotal in sustaining the expedition through the pandemic.

Stefan Schwarze and Thomas Wunderlich, the *Polarstern's* experienced and astute captains, were crucial to her smooth progress. Of course, an undertaking like MOSAiC would have been unthinkable without the incredibly dedicated *Polarstern* crew, who went above and beyond the call of duty. They carried the expedition. I thank all crew members and both captains for their commitment, their impressive performance throughout the expedition, and for the unforgettable time we shared in the Arctic ice. I will never forget this crew, or what they achieved! I also thank the "Deutsche Wetterdienst" (DWD) for their skillful advice on board and from land, and the HeliService pilots and technicians for their relentless operation of the helicopters under most difficult conditions throughout the expedition.

I also thank the captains and crew of all the partner ships who assisted the expedition and provided essential supplies, often in the most challenging ice conditions, as well as those who came to our aid when the pandemic struck: the *Akademik Fedorov, Kapitan Dranitsyn, Admiral Makarov, Maria S. Merian, Sonne,* and *Akademik Tryoshnikov.* We had a wonderful time on board all these ships, and that's down to their crew. I am also grateful to the German Research Fleet Coordination Centre, the German Research Foundation, and the German Federal Ministry of Education and Research for making the *Maria S. Merian* and *Sonne* available at short notice and saving the expedition.

MOSAiC was a complex undertaking, made possible by the amazing efforts of so many people at the Alfred Wegener Institute. I thank my assistant, Sabine Helbig, and all my colleagues for being so helpful and supportive when things were hectic and for taking on so much work; thanks also to everyone in the Controlling, Purchasing, HR, Communication, and Media departments, the general administrative team, the specialist teams, and the board of directors for their tireless commitment to the expedition.

I thank the Arctic and Antarctic Research Institute (AARI) in Saint Petersburg for our long-standing, trusting, and close working relationship—particularly Alexander Makarov, AARI director, and Vladimir Sokolov, head of Arctic operations. MOSAiC wouldn't have been feasible without our Russian friends and colleagues.

Special thanks go to Otmar Wiestler, president of the Helmholtz Association of German Research Centres, for his amazing support and infectious enthusiasm for our research!

I would also like to personally thank Anja Karliczek, German minister of education and research, for supporting the expedition in good times and bad, particularly when the coronavirus brought MOSAiC to the brink of failure.

I am grateful to Heiko Maas, German minister of foreign affairs, for our fascinating discussions on the wider implications of the dramatic climate change taking place in the Arctic, particularly its impact on international relations and preventive conflict management. These conversations have been a source of inspiration and motivation. I also thank the German Federal Foreign Office for helping our international team members to travel during the pandemic.

Thanks to the International Arctic Science Committee (IASC) and Volker Rachold, who led the committee at such an important time. IASC played a key role in establishing the international collaboration so crucial to MOSAiC.

MOSAiC is the sum of its partners' contributions. I would like to expressly thank each of the more than eighty partner institutions, funding agencies, and scientific institutes from twenty different countries.

Thank you to Marlene Göring for revising the text, filling in gaps, and creating many of the info boxes, to Arno Matschiner for editing the text, and to Karen Guddas and everyone at C. Bertelsmann for their amazing support in making the book happen.

The expedition thrived on its participants. They are the ones who made it a success, who collected the crucial data and samples and made MOSAiC an unforgettable experience for all involved. I thank all the expedition members for the fantastic year we spent together, a year that will change science forever. I thank Christian Haas and Torsten Kanzow, who each coordinated an expedition phase, and the leaders of the Atmo, Ice, Ocean, Eco, BGC, Logistics, Modeling, Data, Communication, Remote Sensing, and Aircraft teams for their tireless efforts to implement our program. Finally, I thank the families and friends of all the participants for putting up with the long absences and helping us through tough times with encouraging messages from home.

NOTES

1. Fridtjof Nansen, *Farthest North. Being the Record of a Voyage of Exploration of the Ship "Fram" 1893–1896*, vol. 1 (New York and London: Harper & Brothers, 1897; Project Gutenberg, 2009), 1–2, gutenberg.org/files/30197/30197-h/30197-h.htm.

2. Nansen, *Farthest North*, 1:577.

3. Fridtjof Nansen, *Farthest North. Being the Record of a Voyage of Exploration of the Ship "Fram" 1893–1896*, vol. 2 (New York and London: Harper & Brothers, 1897; Project Gutenberg, 2010), 318, gutenberg.org/files/34120/34120-h/34120-h.htm.

4. Laura Backes and Raphael Thelen, "We Are Engaged in the Murder of the World's Children," *Spiegel International*, November 22, 2019, spiegel.de/international/europe/interview-with-extinction-rebellion-co-founder-roger-hallam-a-1297789.html.

IMAGE CREDITS

All images by Markus Rex with the exception of:
vi–vii (map): Peter Palm
6–7: Alfred Wegener Institute / Martin Kuensting (CC-BY 4.0) and Tim Wehrmann
9: Tim Wehrmann
10 top: National Library of Norway
10 bottom: National Library of Norway / Henry Van der Weyde
20, 112, 125 top left and top right, plate 1 bottom: Alfred Wegener Institute / Stefan Hendricks (CC-BY 4.0)
39, 58: Alfred Wegener Institute / Esther Horvath (CC-BY 4.0)
125 bottom: Alfred Wegener Institute / Marcel Nicolaus (CC-BY 4.0)
227, 229: Alfred Wegener Institute / Steffen Graupner (CC-BY 4.0)
234, 240: Alfred Wegener Institute / Lianna Nixon (CC-BY 4.0)

INDEX

Fram Strait, 11, 196, 213, 220, 227
Frank, Gerd, 102
Franklin expedition, 138
Franz Josef Land, 131, 142, 146, 180, 182
frostbite, 89, 254

Gakkel Deep, 22
GEM, 24, 37
geostationary satellites, 118
Germany, 156, 157, 158, 164–65, 176, 217. See
    also *Maria S. Merian*; *Sonne*
global warming. *See* climate change
Grafe, Jens, 102
Greenland, 9, 212–13, 231, 232–33, 255. *See also*
    Station Nord
Grundmann, Uwe, 102

Haapala, Jari, 145
Haas, Christian, 123
Hallam, Roger, 257
Hannula, Henna-Reetta, 240, 241
Heitland, Tim, 170
helicopters: for driving away polar bears, 63,
    68, 207, 209; emergency operational plans
    and, 126; for exploration, 23, 27–28, 28;
    goodbyes to *Fedorov* and, 73; landing pad
    on ice floe, 54–55; rescuing ROV City, 78;
    survival suits and, 22; weather reports and,
    192. *See also* airplanes
Heuck, Hinnerk, 97
Hildebrandt, Nicole, 82
Honold, Hans, 92, 108, 119–20
Horvath, Esther, 77

ice: about, 30; breaking process, 213–14,
    220–21; climate change and, 35–36, 84,
    182–83; core samples, 82, 97; dark nilas,
    29; dynamic secondary ice growth, 83,
    237; finger rafting, 29, 30; freeze-thaw
    cycle, 82–84; improvised bridges and
    ferries for crossing, 112, 112, 212, 213, 240,
    240; initial freeze, 29, 245–46; light nilas,
    29; mountains formed from, 72, 103, 115,
    126–27; Nansen on, 182; one-year vs.
    multiyear ice, 83–84; pancake ice, 29, 30;
    pressure from, 69, 70–71, 76–77, 103, 122,
    141–42, 203–4; pressure ridges, 18, 63, 72,

72, 103–4, 104, 114, 250; pushing through,
    19–21, 20, 24, 153; reading, 119; research on,
    120, 249–50; retreating, 35–36; rotation,
    220; satellite surveillance, 27; sediment in,
    186–87, 189, 212; slush ice, 29, 30; summer
    instability, 24–25; summer thaw, 83–84,
    191–93, 209–10, 210, 212, 213, 214, 234–35,
    236, plate 4; thermodynamic freezing, 83;
    Transpolar Drift and, 210; travel methods
    for, 110; working conditions on, 125. *See
    also* MOSAiC ice floe and research camp
    1.0; MOSAiC ice floe and research camp 2.0
ice anchors, 76, 185
iceblink, 19
icebreakers, 176, 204
ice picks, 117
Indigenous peoples, 217
Inuit, 217–19
Is Fjord, 172, 174
Italy, 156–57

Jeannette (ship), xi, 11
jet stream, 152, 249
Johansen, Hjalmar, 10, 11, 179–80

Kanzow, Torsten, 123
Kapitan Dranitsyn (Russian icebreaker):
    Christmas aboard, 143–45; COVID-19
    pandemic and, 157–58; exchanging
    supplies and personnel with, 123, 139–40;
    first journey to *Polarstern*, 122–23, 127,
    128, 131, 132, 133–34, 137–39; first return
    to Tromsø, 140–41, 141–42, 143, 145–46,
    146–47, plate 4; fuel transfer from *Makarov*,
    154–56; *Lance* rescue effort and, 142;
    preparation for arrival of, 132; second
    journey to *Polarstern*, 152–54; second
    return to Tromsø, 156; support role for
    Polarstern, 94, 176
Kara Sea, 8, 12, 17–18, 209
Kara Strait, 8
Karliczek, Anja, 1, 177
kayaks, 109, 110, 179
kittiwakes, 178–79, 179
Kohlberg, Eberhard, 170
Kohltour, 101, 124
Kongs Fjord, 32, 33–35, 35, 60

DAVID SUZUKI INSTITUTE

THE DAVID SUZUKI INSTITUTE is a non-profit organization founded in 2010 to stimulate debate and action on environmental issues. The Institute and the David Suzuki Foundation both work to advance awareness of environmental issues important to all Canadians.

We invite you to support the activities of the Institute. For more information please contact us at:

David Suzuki Institute
219 – 2211 West 4th Avenue
Vancouver, BC, Canada V6K 4S2
info@davidsuzukiinstitute.org
604-742-2899
www.davidsuzukiinstitute.org

Cheques can be made payable to The David Suzuki Institute.